P9-AFF-209
WITHDRAWN
UTSA LIBRARIES

Engineering Polymer Sourcebook

Engineering Polymer Sourcebook

Dr. Raymond B. Seymour
Distinguished Professor,
Department of Polymer Science,
The University of Southern Mississippi

McGraw-Hill Publishing Company
New York St. Louis San Francisco Auckland Bogotá
Caracas Hamburg Lisbon London Madrid Mexico
Milan Montreal New Delhi Oklahoma City
Paris San Juan São Paulo Singapore
Sydney Tokyo Toronto

Library of Congress Cataloging-in-Publication Data

Seymour, Raymond Benedict,
Engineering polymer sourcebook / Raymond B. Seymour.
p. cm.
Includes bibliographies and index.
ISBN 0-07-056360-8
1. Polymers. I. Title.
TP156.P6S44 1990
668.9—dc20 89-35066
CIP

1234567890 DOC/DOC 89432109

ISBN 0-07-056360-8

The editors for this book were Robert W. Hauserman and Lucy Mullins, the designer was Naomi Auerbach, and the production supervisor was Richard A. Ausburn. It was set in Century Schoolbook. It was composed by the McGraw-Hill Publishing Company Professional and Reference Division composition unit.

Printed and bound by R. R. Donnelley & Sons Company.

Contents

Foreword

The advancement of polymers as materials of construction has been dependent, to a large extent, on the synthesis and modification of many different high-performance polymers which have been described in many outstanding books on this subject.

However, since many design engineers have not been trained in the field of polymer science and technology, their attempts to choose and use polymers are hampered by the absence of engineering-oriented polymer science reference books.

Fortunately, Professor Seymour has helped to fill this void by authoring this *Engineering Polymer Sourcebook*. I am impressed by the scope of the contents of this book and recommend it to those who are seeking more design information on these essential high-performance materials.

Herman F. Mark

Preface

Although they may not have been used for engineering applications, such as gears and structural members, many polymers used prior to World War II were low-grade engineering polymers. If we exclude strong, natural polymers, such as wood and cotton, the most noteworthy ancient derivative of natural polymers was leather, which has been called "the most historic of useful materials." The primitive techniques used several thousand years ago have been improved, but the modern tanning process is similar to that used by our ancestors a few millennia ago.

It has been recorded that bitumens, which were called slime, were used as caulking materials to waterproof Noah's Ark, Moses' basket, and ancient water tanks. These low-grade engineering applications may be overlooked today, but they were essential for survival of the inhabitants several thousand years ago.

Likewise, the vulcanization of rubber for the production of flexible and hard rubber and the plasticization (flexibilization) of cellulose nitrate for the production of moldable celluloid, cellulose nitrate coatings, and artificial fibers in the nineteenth century may not be considered engineering "breakthroughs," but these products were essential for waterproofing textiles, protecting metal surfaces, providing molded articles, and for producing continuous filaments.

Some polymer technologists exclude thermosets in their list of engineering polymers. However, Glyptal, which was produced by the esterification of glycerol by phthalic anhydride, and Bakelite, which was produced by the reaction of phenol and formaldehyde, are thermosets which can definitely be classified as engineering polymers. Hence, both high-performance thermosets and high-performance thermoplastics will continue to have engineering applications and are discussed in this book.

Several good books, which emphasize the chemistry of engineering polymers, have been published. These are listed as references, and, of course, some chemistry is included in this book. However, the emphasis is on design, processing, fabrication, and application of these unique modern materials of construction.

Raymond B. Seymour

Acknowledgment

The typing services of Mrs. Debi Sones, who typed this book, are gratefully acknowledged.

Chapter

1

History of the Development of Engineering Polymers

1.1 Introduction—Scope

This book concerns engineering polymers, which may be thermoplastic or thermosetting polymers that maintain their dimensional stability and their major mechanical properties in the temperature range of 0 to 100°C.

Other criteria for engineering polymers include moldability and a good balance of mechanical properties which is maintained for a long time over a wide range of environmental conditions. These polymers, which may be natural or synthetic, are the counterparts of traditional engineering materials, such as metals and ceramics.

The term "engineering polymers," which is interchangeable with the terms "high-performance polymers" and "engineering plastics," is subject to a variety of other criteria and interpretations. The editorial steering committee of the American Society for Metals (ASM International) has defined the term "engineering plastics" as a synthetic polymer resin-based material that has load-bearing characteristics and high-performance properties which permit it to be used in the same manner as metals and ceramics. However, contrary to the ASM definition, some natural polymers are included in this book on engineering polymers. Engineering plastics have also been defined as plastics which lend themselves for use in engineering design, such as gears and structural members.

The major products produced by the polymer industry are general-purpose plastics, such as polyethylene, polypropylene, polyvinyl chloride, and polystyrene. Because of their relatively low strength and lack of resistance to moderately high temperatures, these widely used plastics are not classified as engineering polymers. However, some copolymers of styrene do meet the criteria for low-grade engineering polymers.

1.2 Natural Polymers

High-performance polymers are not new. Nature has been generous in supplying us with abundant quantities of tropocollagen, which is the major component of skin, tendons, cartilage, bone, and teeth. Keratin, which is found in hooves, nails, and claws of animals, and in beaks of birds, and fibroin, which is the principal constituent of silk, have also served as moderately high performance polymers for thousands of years. Also, nucleoproteins, which are not defined as high-performance polymers, are essential for our very existence.

Wood, which is a composite of cellulose and lignin, is also a moderately high performance material. Fire, which ranks next to the wheel as humankind's most significant discovery, would have had little impact unless wood was available as fuel. A century ago, wood was still our major source of energy, and this renewable resource continues to be an important structural material.

There are very few other materials with the characteristic elasticity of natural rubber (NR; *Hevea braziliensis*); hence, natural rubber should be classified as a high-performance polymer. Since many of today's high-performance polymers are blends, it is of interest to note that a blend of cis hevea rubber and trans gutta-percha plastic was patented and used as a cable coating in 1846. [The cis and trans prefixes describe the position of chain extensions on the ethylenic double bonds $\left(CH{=}CHZ \right)$ in polymeric dienes. Dienes contain two double bonds.]

1.3 Pre-World War I Polymers

Nelson Goodyear, a brother of Charles Goodyear, patented hard rubber (ebonite) in 1851. This pioneer inventor's manufactured thermoset was used for dental prosthetics, combs, and battery cases. (A thermosetting plastic is one that does not dissolve in solvents and is not softened when heated.) Cellulose nitrate, plasticized, i.e., flexibilized, by camphor, was the pioneer manufactured thermoplastic (Celluloid). It was patented by two brothers, I. S. and J. W. Hyatt, in 1870. (In contrast to the thermosets, thermoplastics are soluble in selected solvents and may be heated and cooled, reversibly, without decomposition.) Few of the early applications of cellulose nitrate were structural, but its subsequent use as an automotive lacquer (Duco) is a relatively high performance application.

Polyesters, which had potential as high-performance polymers, were synthesized by Berthelot in 1853 and commercialized as Glyptal resins by Watson Smith in 1907. General Electric* used this polyester as a heat-resistant coating and applied for several patents on this unique material. (Esters are obtained by the reaction of alcohols and organic acids.)

Some technologists may not classify optically clear polymethyl methacrylate (PMMA) as a high-performance polymer, but it was less brittle than any other available transparent solid when Otto Rohm investigated and patented this polymer in the early 1900s.

Leo Baekeland synthesized, patented, and produced one of the first synthetic thermosets by the condensation of phenol and formaldehyde in 1907. Bakelite was the first commercially available engineering polymer.

The first synthetic elastomer (methyl rubber) was produced by the polymerization of dimethylbutadiene by I. Kondakov in 1900. During World War I, 2000 metric tons (t) of this elastomer was produced by the German chemical industry. This polymer would have provided superior performance if it had been reinforced by carbon black. Fortunately for the Allied forces, the Germans were not aware of the proprietary use of carbon black as a reinforcing filler in English Silvertown cord tires. The addition of 60 parts of carbon black to 40 parts of methyl rubber increases its tensile strength by at least 100 percent.

In addition to these pioneer polymeric coatings, thermosets, glazings, and elastomers, cellulose diacetate fibers were made by G. W. Miles, who partially saponified (hydrolyzed) acetone-insoluble cellulose triacetate to produce a polymer which was soluble in inexpensive solvents, such as acetone.

The relatively high performance character of cellulose diacetate was demonstrated during World War II, when it was used successfully as a coating (dope) for fabric on British airplanes. Since it is soluble in solvents and is reversibly softened by heat, cellulose acetate is a good illustration of a thermoplastic.

1.4 Pre-World War II Polymers

In spite of the commercialization and subsequent unprecedented growth of the three major general-purpose thermoplastics, namely, polyvinyl chloride (PVC), polystyrene (PS), and polyethylene (PE), which do not meet the criteria for engineering polymers, several relatively high performance polymers introduced in the 1920s and 1930s captured a share of the polymer market and are still in use today. J. C. Patrick patented a solvent-resistant polyethylene sulfide (Thiokol),

*See the appendix for full company names.

and E. Konrad and R. E. Tschunkur patented the heat-resistant elastomeric copolymer of butadiene and acrylonitrile (NBR; Buna N) in the late 1920s.

W. H. Carothers and coworkers synthesized a heat- and solvent-resistant polychlorobutadiene (neoprene); R. W. Thomas and W. Sparks patented a permeation-resistant copolymer of isobutylene and isoprene (butyl rubber); and E. Rochow commercialized high-heat- and moisture-resistant silicones in the 1930s.

R. Kienle upgraded the Glyptal polymers by the incorporation of unsaturated vegetable oils. These condensation polymers (alkyds) became the major coating resins and paved the way for the commercialization of polyester fibers (PET) by J. R. Whinfield and J. T. Dickson and of reinforced unsaturated polyesters by Ellis in the late 1930s. Another thermoset (epoxy resin), which was characterized by good adhesion to metallic surfaces and excellent resistance to alkalies, was patented by P. Schlack in 1938.

W. H. Carothers and J. W. Hill of Du Pont produced aliphatic polyesters but shelved further investigation of these substances in favor of high-performance polyamides (nylon). O. Bayer produced competitive fibers from polyurethanes, but these high-performance polymers were used primarily for foams, molded plastics, and coatings. R. Plunkett also produced polytetrafluoroethylene (PTFE; Teflon), which is one of the most corrosion-resistant products available.

1.5 Engineering Thermoplastics

The "big five" engineering thermoplastics, with their annual volume (1987) in thousands of metric tons in the United States and the world, shown respectively in parentheses, are as follows: nylons (214) (605), polycarbonate (176) (306), acetals (55) (246), polyphenylene ether (80) (158), and thermoplastic polyesters (55) (126).

Acrylics, epoxies, phenolics, and unsaturated polyesters are all high-performance polymers and are used in the United States at an annual rate (1987) of 302,000, 184,000, 1,260,000, and 598,000 t, respectively. The development of these polymers has been described in several books which are listed as references at the end of this chapter.

1.6 Nylons

The story of the development of nylon 6,6 is a case history of unprecedented modern management of industrial fundamental research. In 1927, C. M. A. Stine, director of the Du Pont chemical department, requested a modest budget for fundamental research and in 1928 hired Wallace H. Carothers and a team of six chemists to conduct these investigations.

The first useful products of this research were aliphatic polyesters produced in the early 1930s by the condensation of a difunctional acid [adipic acid: $HOOC(CH_2)_4COOH$] and a difunctional alcohol (ethylene glycol: $HOCH_2CH_2OH$). [Aliphatic polymers consist of continuous chains of carbon atoms, and their structure is similar to that of gasoline or paraffin wax. All organic acids contain a carboxylic group (COOH), and all alcohols contain a hydroxyl group (OH). There are two hydroxyl groups in a glycol.] The melting points of these polyesters were below those required for general-purpose polymers. However, this research set the stage for the subsequent production of aromatic polyesters and the synthesis of polyamides which do meet the requirements of engineering polymers. (Aromatic polymers contain benzene rings which stiffen and increase the temperature resistance of polymers.)

High-molecular-weight polyamides, originally called "superpolyamides," were produced by Carothers and Julian Hill in 1935 by the condensation of a difunctional acid (adipic acid) and a difunctional amine: hexamethylenediamine [$H_2N(CH_2)_6NH_2$]. Fibers for stockings and bristles for brushes were produced from nylon 6,6 in a pilot plant in Arlington, New Jersey, in 1937 and 1938, respectively.

The first commercial nylon 6,6 fiber was produced in a pilot plant in Seaford, Delaware, in 1940, and the first molding powder, for the fabrication of engineering plastics, was made available to Du Pont customers in 1941. Nevertheless, the use of nylon 6,6 as an engineering polymer was delayed because of the need for nylon 6,6 filaments for parachutes, etc., during the war and, after the cessation of hostilities, because of the slow development of modern injection-molding presses capable of molding this high-melting-point (265°C) polymer. The melting point of nylons (PA) is directly related to the concentration of carbonyl (CO) groups and inversely to the concentration of methylene groups (CH_2) in the polymer. Nylon is more readily molded in the screw-type injection-molding press, which was introduced in 1956, than in the classic plunger-type molding press. In the injection-molding process, a heated polymer is forced into a cooled two-piece mold and the molded part is ejected after it solidifies. The equations for the synthesis of nylon 6,6 are:

$$MHOOC(CH_2)_4COOH + MH_2N\,(CH_2)_6NH_2 \rightarrow$$

$$MHOOC(CH_2)_4COO^-,\ {}^+H_3N(CH_2)_6NH_2$$

Nylon 6,6 salt

$$MHOOC(CH_2)_4COO^-,\ {}^+H_3N(CH_2)_6NH_2 \xrightarrow{\Delta\text{(heat)}}$$

$$MH_2O + \mathrm{+}OC(CH_2)_4CONH(CH_2)_6NH\mathrm{+}$$

Nylon 6,6

Polymers may be classified on the basis of their end use as elastomers, plastics, fibers, coatings, adhesives, foams, etc. These are all high-molecular-weight products which differ in their degree of crystallinity and in the strength of their intermolecular forces. Unstretched elastomers (rubbers) are noncrystalline (amorphous) polymers which have weak (London) intermolecular forces. In contrast, most fibers have a high degree of crystallinity and have strong (hydrogen bonds) intermolecular forces. The structural properties of plastics, which may be used as coatings, adhesives, or foams, are in between these two extremes. The intermolecular forces, ranging from 2 kilocalories (Kcal) for the weak London forces present in natural rubber to about 10 Kcal for the strong hydrogen bonds present in nylon and cellulose, are present in most fibers and engineering plastics.

Nylon 6,6 is diadic, i.e., made from two reactants. Other diadic nylons, which are also called the AABB type, are made by the condensation of higher-molecular-weight dicarboxylic acids, such as sebacic acid [$HOOC(CH_2)_8COOH$] and have lower melting points than nylon 6,6. Thus, nylon 6,10 has a melting point of 220°C. The melting points of monadic nylons, i.e., those made from one reactant, such as nylon 6 $\text{+}OC(CH_2)_5NH\text{+}$, are also lower than that of nylon 6,6. Paul Schlack produced nylon 6 (MP 226°C) in 1938 by the alkaline-catalyzed polymerization of a cyclic compound called caprolactam.

The production of large parts, such as bumpers and tanks, from nylon 6 (Nyrim) has been made possible by the incorporation of activators (acetylated compounds) for the polymerization of caprolactam in situ in what is called the "monomer casting process." Lower-melting-point commercial polyamides, such as nylon 11 (MP 180°C) and nylon 12 (MP 179°C), were developed by Joseph Zeltner in 1938 and J. Dachs and E. Schwartz in 1962. These polymers, which are more resistant to moisture than nylon 6,6 or nylon 6, are used, to a small extent, in engineering applications.

Du Pont and Monsanto also produce mineral-filled and fiberglass-reinforced nylon 6,6 under the trade names Zytel and Vyndyne, respectively. Long-fiber-reinforced nylon 6,6 (Verton), which has a much higher impact resistance than short-fiber-filled nylon 6,6, is being produced by pultrusion techniques, in which continuous filaments, impregnated with polymer, are pulled through a heated dye. Several books on the history of nylon and other high-performance polymers are listed at the end of this chapter.

1.7 Polyesters

While the volume of nylon 6,6 exceeds that of all other engineering polymers, the variety of available polyesters exceeds that of the polyamides and all other polymers. The original commercial polyesters were the thermosetting Glyptals, which were produced by the condensation of glycerol [$H_2C(OH)CH(OH)CH_2OH$] and phthalic anhydride ($C_6H_4C_2O_3$) by Watson Smith in 1903. [An anhydride (C_2O_3) is produced by the dehydration of two COOH groups.] These thermally curable polyesters were modified by R. Kienle, who produced oleoresinous-like unsaturated polyesters in 1933 by the condensation of ethylene glycol, phthalic anhydride, and unsaturated vegetable oils, such as those containing linolenic acid ($C_{17}H_{29}COOH$).

The condensation of bifunctional phthalic anhydride and bifunctional ethylene glycol produces a thermoplastic polyester—polyethylene phthalate, $-\!\!\left[O_2CC_6H_4CO_2(CH)_2\right]\!\!-$. The molecular weight of these and other linear condensation polymers is related directly to the purity of the reactants. Thermosets or three-dimensional polymers can be produced when a trifunctional reactant, such as glycerol, is used. The third hydroxyl group in glycerol is less reactive than the two primary hydroxyl groups, and thus the curing or cross-linking of Glyptals can be controlled by the application of heat.

The alkyds are linear (continuous chain) polymers which contain unsaturated ethylenic $-\!\!(CH{=}CH)\!\!-$ groups in the polymer chain or backbone. These unsaturated groups may undergo another type of polymerization, i.e., addition or chain polymerization, to produce infusible, cross-linked polymers. (The two principal polymerization reactions are condensation (step reaction), which is used to produce nylon or polyesters, and addition (chain reaction), which is used to produce PS, PVC, and LDPE. A reactive species, such as an organic peroxide, is required for the initiation of the chain reaction in addition polymerization.) The curing of alkyds is similar to the curing (drying) of classic oleoresinous paints. The curing in the air may be catalyzed or accelerated by the incorporation of heavy metal salts of organic acids (driers), such as the cobalt ester of linoleic acid.

In 1937, C. Ellis and J. B. Rust produced an unsaturated linear polymer (polyethylene maleate) by the condensation of maleic anhydride [$(CHCO)_2O$] and ethylene glycol. This prepolymer can be dissolved in styrene monomer ($C_6H_5CH{=}CH_2$), and then the components of this solution can be polymerized by addition or radical chain polymerization in the presence of an organic peroxide to produce a thermoset polyester. These polymers, which are usually reinforced by fibrous glass, are called "unsaturated polyesters," simply "polyesters," or even "glass" such as a "glass boat."

Esters of allyl alcohol ($H_2C{=}CHCH_2OH$) and dicarboxylic acids, such as phthalic acid [$C_6H_4(COOH)_2$] were polymerized by B. S. Garvey and C. H. Alexander in 1940 to produce clear thermoset polymers (Dapon). A comparable thermoset allyl polymer (CR-39; DADC) is produced by the organic-peroxide-catalyzed polymerization of diethylene glycol bis(allyl carbonate).

Thermoplastic polymers of esters of acrylic ($CH_2{=}CHCOOH$) and methacrylic acid [$CH_2{=}C(CH_3)COOH$] were developed by Otto Rohm in 1901. These clear polymers (PMMA), which are sold under the trade names Plexiglas and Lucite, serve a unique purpose as optically clear, impact-resistant structural materials. The esters of α-cyanoacrylate [$H_2C{=}C(CN)COOR$], such as butyl α-cyanoacrylate [$H_2C{=}C(CN)COOC_4H_9$], polymerize readily in air to produce exceptionally strong adhesives (Krazy Glue).

Carothers made aliphatic polyesters in 1936 by the condensation of adipic acid and ethylene glycol. Comparable polyesters which meet the specifications for engineering polymers were produced in 1940 by J. R. Whinfield and J. T. Dickson by using terephthalic acid ($HOOCC_6H_4COOH$) instead of adipic acid. This high-melting-point (264°C) crystalline polyester (Terylene) was used principally for the production of polyester fibers such as Dacron.

This polyester (PET) could not be used as an engineering polymer because of slow crystallization in conventional water-heated molds. However, this deficiency was overcome, to some extent, by the use of hotter molds (130°C) and by annealing the molded parts.

In 1978, T. J. Deyrup of Du Pont produced a useful polyester engineering polymer (Rynite) by a two-step crystallization system. This system includes chemical nucleation by sodium ions, such as those from the sodium salt of ionomers (copolymers of ethylene and methacrylic acid), and the incorporation of an additive which increases polymer chain mobility by lowering the glass transition temperature. (The glass transition temperature T_g, which has been called the second-order transition temperature, is the temperature at which wriggling of the polymer chain occurs as the temperature is raised, i.e., the polymer changes from a glass to a flexible solid. The melting point T_m is the first-order transition at which the crystals and the molten polymer are in equilibrium.) The first additive of this type was the dibenzoate ($C_6H_5COO{-}$) of neopentyl glycol [$HO(CH_2)_5OH$]. This modification permitted the use of conventional water-heated molds.

Polybutylene terephthalate {PBT; $\mathrm{+O_2CC_6H_4CO_2(CH_2)_4+}$} which is a more readily crystallizable, high-melting-point (223°C) polyester, was introduced commercially in 1969 by Celanese (which is now called Hoechst-Celanese). Over 100,000 t of fiberglass-reinforced PET and PBT was injection-molded in 1981, and it is anticipated that this annual volume will triple by 1992.

Several investigators produced amorphous aromatic polyesters by the condensation of an aromatic glycol, such as bisphenol A [$HOC_6H_4C(CH_3)_2C_6H_4OH$], and phthaloyl dichloride in the early 1960s. [Amorphous polymers, which are noncrystalline materials, are characterized by their T_g. Crystalline polymers which contain crystalline and some amorphous regions are characterized by their T_m. Clear polymers, such as polycarbonate (Lexan), are amorphous. Opaque polymers, such as nylon 6,6, contain both crystalline and amorphous domains.] In 1974, Uniticka produced a commercial polyarylate (U-polymer), and Union Carbide introduced Ardel and Durel in 1978; these arylate polymers are sold under these trade names by Amoco and Hoechst-Celanese.

In 1956, Herman Schnell of Farbenfabricken Bayer and D. W. Fox of General Electric independently produced commercial polycarbonates by the condensation of bisphenol A and phosgene ($COCl_2$). This amorphous, transparent, impact-resistant engineering polymer is sold under the trade names Lexan, Merlon, and Calibre by General Electric, Bayer, and Dow, respectively. Over 140,000 t of polycarbonate was produced in the United States in 1987.

1.8 Polyacetals

While polyoxymethylene (POM; polyformaldehyde), which is produced by the polymerization of formaldehyde (CH_2O), had been known for many decades, it was not commercially available until 1952, when R. N. MacDonald of Du Pont stabilized this polymer by capping the hydroxyl end groups by acetyl groups. The uncapped polymer is readily decomposed by heat. The acetyl-stabilized polyacetal [Delrin; $H_3CCO(CH_2O)_nOCCH_3$] is a tough, heat-resistant crystalline engineering polymer.

In 1961, Celanese produced a thermally stable copolymer of formaldehyde and ethylene oxide

$$\begin{array}{c} H_2C - CH_2 \\ \diagdown \;\; \diagup \\ O \end{array}$$

(Celcon). Over 50,000 t of acetal resins were produced in the United States in 1987.

1.9 Polysulfones

In the early 1960s, J. B. Rose of ICI and A. G. Farnham and R. N. Johnson of Union Carbide produced transparent, yellow polysulfones, under the trade names Victrex and Udel, respectively. In 1967, 3M introduced another polysulfone under the trade name Astrel, and Union Carbide introduced related polymers (Radel) with sulfone (SO_2) stiffening groups in the polymer chain. These commercial polymers differ slightly in chemical structure and in the technique used for their production.

1.10 Polysulfides

Polysulfide elastomers (Thiokol) were produced in 1927 by J. C. Patrick by the condensation of sodium tetrasulfide (Na_2S_4) and ethylene dichloride [$Cl(CH_2)_2Cl$]. Because of their unique solvent and temperature resistance, these aliphatic elastomers are classified as engineering polymers. They continue to be produced in small quantities, but liquid Thiokol LP-3, which is obtained by the reduction of the solid elastomer, is widely used as a binder in solid propellants and as a caulking and sealing material. This liquid product, which was also discovered by Patrick, is converted to the solid elastomer in situ by the addition of an oxidizing agent, such as lead oxide.

An aromatic polyethylene sulfide was produced commercially in 1967 by J. T. Edmunds and H. W. Hill by the condensation of sodium sulfide with *p*-dichlorobenzene (ClC_6H_4Cl). This solvent and temperature-resistant engineering polymer is produced by Phillips Petroleum and Hoechst-Celanese under the trade names Ryton and Fortron, respectively.

1.11 PEEK

Since the carbonyl group (C=O), like the sulfonyl group (SO_2), is a stiffening group, J. Rose produced an amorphous polyarylether ketone (PEEK) in 1975, using a procedure similar to that used for making polyarylsulfones. The commercial PEEK [T_m = 343°C (621°F)] is produced by the condensation of the potassium salt of bisphenol A [$KOC_6H_4C(CH_3)_2C_6H_4OK$] and bis-4-fluorophenyl ketone ($FC_6H_4COC_6H_4F$).

1.12 PPO

In 1965, A. Hay of General Electric produced polyphenylene oxide (PPO) by the copper-catalyzed oxidation of 2,6-xylenol [$(CH_3)_2C_6H_3OH$]. In spite of the flexibilizing effect of the ether group (O), this high-melting-point (mp = 260°C) crystalline polymer was difficult to process. This difficulty was overcome by blending with PS: $-[CH_2CH(C_6H_5)]-$. The blend (Noryl), which is also called polyphenylene ether, has a T_g which is between that of PS and PPO and is related to the ratio of constituents in the commercial polymeric blend. In 1987, 70,000 t of PPO was produced in the United States.

1.13 Polymeric Blends

The success of the Noryl blend catalyzed the investigation of many other blends, such as Xenoy, a blend of polycarbonate (PC) and PBT, by General Electric; PC and acrylonitrile-butadiene-styrene terpolymer (ABS) under the trade name Bayblend by Mobay; ABS and polysulfone blends (Mindel) by Union Carbide; nylon and polyethylene (Zytel) by Du Pont; PC and styrene-maleic anhydride (SMA) copolymer (Arloy) by Arco; polyacetal and elastomer (Delrin) by Du Pont; PET and elastomer by Du Pont (Rynite); PBT and PET (Celanex) by Hoechst-Celanese; PPO and SMA by General Electric; and PC and nylon by Dexter.

The overall properties P of blends, or alloys, are equal to the sum of the properties of the components modified by their concentrations C plus a parameter I multiplied by the component properties P_1P_2. In many polymer blends, the value of I approaches zero, i.e., they are nonsynergistic. The value of I is greater than zero for alloys and synergistic mixtures. The equation for calculating the properties of blends or mixtures is

$$P = P_1C_1 + P_2C_2 + IP_1P_2$$

There are over 175 commercial grades of PPO (Noryl, Prevex) available in the United States, and the successful use of these PPO blends has catalyzed the development of many other commercial blends and alloys. The U.S. annual production of polymer alloys in 1987 was 90,000 t and much of this consisted of PPO blends.

ABS (Cycolac), which was one of the first commercial blends, was patented by W. Calvert in 1966. The first products sold by Borg-Warner were blends of NBR and styrene-acrylonitrile copolymer (SAN). These blends have been modified and converted to graft copolymers by some producers to meet the consumer demand for tough, moderately expensive high-impact plastics. In graft copolymers, the chain extension by the comonomers consists of branches on the principal polymer chain.

1.14 Polyimides

There were many U.S. Air Force–funded investigations of high-temperature-resistant polymers in the early post-World War II era. Many of the polymers investigated were based on twined chain backbones and were called ladder polymers. The rationale behind these ladder polymers was that their integrity could be maintained if the bond in one chain was cleaved. Polyimides (PIs) and polybenzimidazoles (PBIs) are now commercially available.

M. T. Bogert and R. R. Renshaw produced PI in 1908 by the thermal dehydration of 4-aminophthalic anhydride ($NH_2C_6H_3C_2O_3$). In the late 1950s, Edwards and Robinson of Du Pont produced PI by a two-step condensation of the dianhydride of an aromatic tetracarboxylic acid ($O_3C_2ArC_2O_3$) and an aromatic diamine (H_2NArNH_2) in a polar solvent. A linear prepolymer, called a polyamic acid, was produced by the partial thermal dehydration in the first step, and this polyamic acid was converted to PI (Skybond, Pyralin) by thermal or chemical imidization in the second step. In the early 1970s, H. R. Lubowitz, in what is now called the polymerization of monomeric reactants (PMR) concept, stabilized the polyamic acid precursor of PI by capping the end groups with derivatives of norbornene anhydride.

The intractability of PI was overcome by the production of polyamideimide (PAI; Torlon) by Amoco and polyetherimide (PEI; Ultem) by Heath and Wirth of General Electric in the 1970s.

PAI is produced by the condensation of trimellitic trichloride [$(ClCO)_3C_6H_3$] and methylene dianiline ($H_2NC_6H_4CH_2C_6H_4NH_2$). This amber-gray, amorphous polymer, which may be injection-molded, has a heat deflection temperature of 274°C.

PEI has been produced by the condensation of an aromatic dianhydride containing ether groups and an aliphatic diamine. This amorphous, injection-moldable polymer has a heat deflection temperature of 200°C.

1.15 Styrene Polymers

PS, which was synthesized by E. Simon in 1839 and produced commercially by Dow in the 1930s, is a clear, brittle amorphous resin which does not meet the requirements of an engineering polymer. However, in the late 1920s, T. Wagner-Jauregg produced a copolymer of styrene and maleic anhydride. This first copolymer (SMA) was useful at higher temperatures than PS but was difficult to mold.

An injection-moldable copolymer of acrylonitrile (AN) ($H_2C{=}CHCN$) and styrene which was produced in Germany in the 1930s also had a higher T_g than PS, and this T_g value was related to the AN content of the copolymer (SAN).

In the early 1940s, R. B. Seymour produced moldable, transparent terpolymers of SMA and AN, which were compounded and sold as engineering polymers by Monsanto under the trade name of Cadon. Arco also introduced moldable engineering polymers based on SMA which were sold under the trade name Dylark.

In the mid-1920s, I. Ostromislensky improved the impact resistance of PS by blending it with NR. A blend of PS and copolymers of styrene and butadiene ($H_2C{=}CHCH{=}CH_2$) called high-impact PS (HIPS) was patented by R. B. Seymour of Monsanto, and blends of PS and styrene-butadiene elastomer (SBR) called Styralloy were produced by R. Boyer of Dow Chemical in the early 1940s.

In the early 1950s, L. E. Daly of Naugatuck Chemical produced blends of SAN and NBR which were called Kralastic. In 1958, W. E.

Calvert produced a high-impact polymer (ABS) which was sold under the trade names Marbon and Cycolac.

1.16 PVC

PVC, which was produced by Regnault in the 1930s, is an amorphous polymer which decomposes at processing temperatures. This difficulty was overcome in the 1930s by E. W. Reid of Carbide and Carbon and R. B. Seymour of Goodyear, who produced moldable copolymers of vinyl chloride ($H_2C{=}CHCl$) and vinyl acetate (Vinylite) and vinylidene chloride ($H_2C{=}CCl_2$) (Pliovic), respectively. W. Semon of Goodrich also produced a flexible PVC by the addition of a high-boiling-point compatible liquid (plasticizer), which he called Koroseal. Nevertheless, the thermal properties of the modified compositions were inferior to those of PVC.

During World War II, chemists at I. G. Farbenindustrie stabilized PVC so that it could be molded and extruded and American chemists improved the heat resistance of PVC by blending with SMA and other elastomeric impact modifiers. About 1,500,000 t of rubber-toughened PVC was used as an engineering polymer in the United States in 1987.

1.17 Fluorocarbon Polymers

In spite of the successful polymerization of vinyl chloride, vinylidene chloride, and their bromine counterparts, most polymer chemists of the early 1930s believed that it was not possible to polymerize vinyl fluoride monomers. However, Roy Plunkett of Du Pont accidentally produced PTFE [$\leftarrow CF_2CF_2 \rightarrow$; Teflon] in 1938 when a solid polymer was discovered in a pressurized tank of gaseous tetrafluoroethylene (TFE). While this heat- and solvent-resistant polymer could not be molded, it could be fabricated by preforming and heating and compressing finely divided PTFE in a process similar to the sintering of metal powders.

The processability of these polyfluorocarbons was improved by the production of a copolymer of TFE and hexafluoropropylene ($CF_3CF{=}CF_2$; FEP) by H. Eleuterio in 1967 and by reducing the number of fluorine atoms on the repeating unit.

Low-molecular-weight polychlorotrifluoroethylene (PCTFE) was patented by I. G. Farbenindustrie in 1957 and commercialized by Hooker Electrochemical during World War II. A higher-molecular-weight engineering polymer (Kel-F) was commercialized by

M. W. Kellogg in the 1950s, and the Kel-F business was acquired by 3M in 1957.

Polyvinylidene fluoride [$\leftarrow CH_2CF_2 \rightarrow$; PVDF; Kynar] was produced by T. A. Ford and W. E. Hanford in 1948 and commercialized by Pennwalt in the early 1960s. Polyvinyl fluoride [$\leftarrow CH_2CHF \rightarrow$; PVF] retains some of the unique chemical resistance of Teflon but is more readily fabricated as a weather-resistant film. Several copolymers of ethylene and fluorocarbon monomers were also produced in the 1960s.

1.18 Phenol Formaldehyde

Phenol-formaldehyde (PF) resin, which is the major thermoset polymer, was commercialized by Leo Baekeland in the early 1900s. The Bakelite resole resins, which were produced by the condensation of phenol (CH_6H_5OH) and formaldehyde under alkaline conditions, were used as linear prepolymers which cured (cross-linked) in the presence of acids or when heated.

The Bakelite novolac resins were thermoplastics which were produced by using an excess of phenol under acid conditions. The formaldehyde required for converting the linear polymer to a thermoplastic in the mold was supplied by the incorporation of hexamethylenetetramine [hexa; $(CH_2)_6N_4$], which releases formaldehyde when heated in the mold.

1.19 Amino Resins

Urea-formaldehyde (UF) plastics were produced by A. Holzer in the late 1800s but were not commercialized until 1908 when H. Goldschmidt, E. W. Neuss, and F. Pollack proposed the use of UF as an adhesive and molding resin. Melamine-formaldehyde (MF) resins, which are similar to UF, were produced in 1935 by Henkel Corp. Some MF is used for electrical applications and molded dishware, but the principal amino plastic is UF.

1.20 Epoxy Resins

Epoxy (EP) resins were produced in 1939 by P. Castans by the condensation of bisphenol A and epichlorohydrin (CH_2OCHCH_2Cl). Commercial EP resins are also produced by the reaction of peracetic acid (H_3CCO_3H) with cycloaliphatic olefins, such as cyclohexane (C_6H_{12}). These prepolymers, which contain EP end groups

(— C —C—)
\ /
O

and pendant hydroxyl groups as the repeating units in the chain, may be cured by the addition of polyamines which react with the EP end groups at room temperature, or by the addition of cyclic anhydrides, which react with the hydroxyl pendant groups at elevated temperatures.

1.21 Cyanate Polymers

In the early 1960s, E. Grigat showed that cyanate esters, such as bisphenol A dicyanate (BADCY), could be processed much like epoxy resins. The heat deflection temperature of BADCY is 255°C. Nobel and Bayer have discontinued the production of cyanate ester engineering polymers, but these are now available from Hoechst-Celanese.

1.22 Polyurethanes

Urethanes produced by the reaction of alkyl isocyanates (RNCO) and aliphatic alcohols (ROH) were synthesized by C. A. Wurtz in 1879, but difunctional reactants were not used until 1937, when O. Bayer obtained polyurethanes by the reaction of alkyl diisocyanates [$R(CNO)_2$] and diols [$R(OH)_2$]. These versatile engineering polymers are used as rigid and flexible foams, fibers, coatings, elastomers, and moldings.

One of the more useful engineering urethane polymers is produced by reaction injection molding (RIM), in which the measured amounts of reactants are injected into a mold cavity where the polymer is produced in situ. Polyureas were also produced by Bayer in the 1940s by the reaction of an alkyl or aryl diisocyanate with an alkyl diamine.

1.23 Silicones

Because of their excellent resistance to elevated temperatures and moisture, silicones [$\text{+}SiR_2O\text{+}$] are widely used as engineering polymers. These inorganic polymers were produced by F. S. Kipping in the early 1900s but were not commercialized until the late 1930s, when Dow Corning, General Electric, and Union Carbide produced them. The thermal resistance of silicones was improved in 1951 by Johanson, who produced polyfluorosilicones.

1.24 Polyphosphazenes

Other inorganic engineering polymers, called polyphosphazenes, $-[N=P(OR)_2]-$, were produced in 1924 by Schenck and Rimer by the thermal polymerization of phosphonitrilic chloride ($NPCl_2$). These engineering polymers, like polyurethanes and silicones, are available as elastomers.

1.25 Elastomers

The first engineering elastomers were produced in 1931 by A. M. Collins and W. H. Carothers. The original polymer was called Duprene [$-(CH_2CCl=CHCH_2)-$], but the name of this heat- and solvent-resistant elastomer was changed to neoprene in the 1950s.

NBR, a copolymer of butadiene and acrylonitrile, was produced in 1930 by E. Konrad and R. E. Tschunkur, who called it Buna N. The oil and heat resistance of this engineering elastomer is related to the AN content of the copolymer. Thiokol and fluorocarbon elastomers are also engineering elastomers. Included in the latter classification are copolymers of TFE and vinylidene fluoride (VF_2) produced by W. E. Hanford and W. Miller in the late 1940s, and VF_2 and hexafluoropropylene (Viton) produced by D. Rexford in the late 1960s.

Another commercial fluorocarbon elastomer is a copolymer of perfluoromethyl vinyl ether ($CF_3OCF=CF_2$) and TFE (Kalrez), which was produced by D. Patterson in the early 1970s. Still another specialty engineering elastomer is Elastopar, which was produced by C. H. Fisher in the late 1940s by the copolymerization of butyl acrylate ($H_2C=CHCOOC_4H_9$) and α-chloroethyl vinyl ether ($ClC_2H_4OCH=CH_2$). Another oil- and heat-resistant elastomer called Hydrin or Parel was produced by E. Vandenberg in the 1960s by the polymerization of epichlorohydrin and propylene oxide.

Butyl rubber, a copolymer of isobutylene [$(CH_3)_2C=CH_2$] and isoprene [$CH_2=C(CH_3)CH=CH_2$], was produced in the late 1930s by W. Sparks and R. Thomas, who copolymerized the monomers by chain polymerization with an aluminum chloride catalyst at extremely low temperatures.

1.26 Thermoplastic Elastomers

Thermoplastic elastomers (TPEs), which contain sequences of hard and soft repeating units in the polymer chain, are readily injection-moldable and do not require cross-linking (vulcanization) to prevent cold flow. TPEs meet the American Society for Testing and Materials (ASTM) definition of an elastic material, i.e., they can be stretched repeatedly to twice their original length and upon the release of the stress, return immediately, with force, to the approximate original length.

In the 1960s, N. Legge of Shell Petroleum produced Kraton, which is a triblock copolymer consisting of sequences of repeating units of styrene-butadiene-styrene, and H. Hsieh of Phillips Petroleum produced Solprene, which is a diblock copolymer of styrene and butadiene. The weatherability of these TPEs has been improved by hydrogenation to produce copolymers which are block copolymers of styrene and butene. The chain extension by the second monomer in a block copolymer is a continuation of the original chain, i.e., A_nB_n.

Bayer produced block copolymers of hydroxyl-terminated polyesters and polyurethanes in the 1940s. Comparable TPEs made from polyethers have better resistance to hydrolysis than those made from polyesters. In 1968, W. Witisiepe produced a copolyester (Hytrel), and in the 1980s other Du Pont and Monsanto high-performance TPEs (Alcryn, Santoprene) were produced from polypropylene (EPDM) and chlorinated ethylene copolymers.

In the 1970s, G. Deleans produced polyether block amide (PEBA) engineering TPEs (Pebax). These versatile polymers can be formulated to meet many different specifications for physical properties, such as high modulus (stiffness).

Block copolymers of polydimethylsiloxane and PS, of polydimethylsiloxane and PC, and of polydimethylsiloxane and polysulfone were commercialized by Dow Corning, General Electric, and Union Carbide in the 1970s.

1.27 Future Outlook

There are only a few producers of high-performance thermoplastics in the United States, namely, Hoechst-Celanese, Du Pont, General Electric, Amoco, Dow Chemical, Mobay, ICI, and Phillips. These firms are expanding their production facilities to meet the anticipated demand of 875,000 t in 1990. The principal end uses will continue to be transportation (34%), electronics (21%), consumer goods (15.5%), industrial applications (10%), and construction (10%). This is now a $1.7 billion business in the United States and should increase to $2.5 billion in 1990.

As shown in Table 1.1, nylon and PC will continue to be the principal high-performance polymers in 1990.

The estimated volume, in metric tons, of some specialty high-performance polymers in 1995 is shown in Table 1.2.

TABLE 1.1 Estimated U.S. Production of High-Performance Polymers in 1990, thousand metric tons

Nylon	260
PC	223
Polyphenylene ether–PS blend	105
PET and PBT	100
Acetals	70
Polyphenylene sulfide	9
Polysulfones	3.5
PEI	3
Liquid-crystal polymers	1.4
Polyether ketones	0.7
PAIs	0.3

TABLE 1.2 Estimated Volume of Specialty High-Performance Polymers, t

High-impact nylon	40
PBT-PC alloy	33
Amorphous nylon	20
Polyimides	16
Polyarylates	14
Polyamide barrier resins	7

Because of the ease of production of blends and alloys on existing equipment, the growth of these high-performance polymers will be on the order of 40 percent in the next decade. Blends account for almost 20 percent of the high-performance thermoplastics marketed at present, and they will retain this percentage.

The worldwide production of high-performance polymers will exceed 2,000,000 t in 1991. The United States, with an annual production of 810,000 t, will continue to lead, but it will be followed closely by western Europe and Japan, with annual volumes of 720,000 and 460,000 t, respectively.

Automobile fenders are already being made from nylon-PPO alloys, and the shift from metal to all-plastic automobile bodies will be a reality in the year 2000. Computer-aided manufacturing (CAM) will catalyze the increased use of polymer blends. Savings of $60,000 for each kilogram reduction in weight of aerospace vehicles favors the use of high-performance polymers.

The development of new high-performance polymers in Japan, West Germany, and the United States will catalyze the use of all types of polymers in a wide variety of applications and open up new opportunities for uses in which satisfactory performance and freedom of design are essential.

The chronological development of commercial polymers is shown in Table 1.3.

TABLE 1.3 Chronological Development of Commercial Polymers

Date	Material
1907	PF resins (Bakelite; Baekeland)
1926	Alkyd polyester (Kienle)
1931	PMMA plastics
1938	Nylon 6,6 fibers (Carothers)
1942	Unsaturated polyester (Ellis and Rust)
1943	Fluorocarbon resins (Teflon; Plunkett)
1943	Silicones
1943	Polyurethanes (Baeyer)
1947	Epoxy resins
1948	Copolymers of AN, butadiene, and styrene (ABS)
1950	Polyester fibers (Whinfield and Dickson)
1956	POM (acetals)
1957	PC
1962	PI resins
1965	PPO
1965	Polysulfone
1970	PBT
1971	Polyphenylene sulfide
1978	Polyarylate (Ardel)
1979	PET-PC blends (Xenoy)
1981	Polyether block amides (Pebax)
1982	PEEK
1983	Polyetheramide Ultem
1984	Liquid-crystal polymers (Xydar)
1985	Liquid-crystal polymers (Vectra)
1986	Polyacetal-elastomer blends (Delrin)
1987	PET-elastomer blend (Rynde)
1987	Engineering polymer
1988	PVC-SMA blend

The thermal properties of typical engineering polymers are shown in Table 1.4.

The titles of typical Underwriters' Laboratory (UL) standards and ASTM tests on plastics are shown in Table 1.5.

TABLE 1.4 Thermal Properties of Engineering Polymers

	Heat-deflection temperature at 1.82 MPa (264 lb/in^2)		UL index*		Thermal conductivity†	
	°C	°F	°C	°F	W/ (m^2• K)	Btu•in/ (h•ft^2•°F)
ABS	99	210	60	140	0.27	1.9
ABS-PC alloy	115	240	60	140	0.25	1.7
Diallyl phthalate (DAP)	285	545	130	265	0.36	2.5
POM	136	275	85	185	0.37	2.6
PMMA	92	200	90	195	0.19	1.3
Polyarylate (PAR)	155	310	—	—	0.22	1.5
Liquid-crystal polymer (LCP)	311	590	220	430	—	—
MF	183	360	130	265	0.42	2.9
Nylon 6	65	150	75	165	0.23	1.6
Nylon 6,6	90	195	75	165	0.25	
Amorphous nylon 12	140	285	65	150	0.25	
Polyaryl ether (PAE)	160	320	160	320	—	—
PBT	—	—	120	250	—	—
PC	129	265	115	240	0.20	
PBT-PC	129	265	105	220	—	—
PEEK	—	—	250	480	0.25	1.7
PEI	210	410	170	340	0.22	1.5
Polyether sulfone (PES)	203	395	170	340	—	—
PET	224	435	140	285	0.17	
PF	163	325	150	300	0.25	1.7
Unsaturated polyester (UP)	279	535	130	265	0.12	0.8
Modified PPO	100	212	80	175	—	—
Polyphenylene sulfide (PPS)	260	500	200	390	0.17	
Polysulfone (PSU)	174	345	140	285	0.26	1.8
SMA terpolymer	103	215	80	175	—	—

*UL index: Temperature at which property values decrease to less than 50 percent at initial values.

†K: kelvin (absolute) temperature, where 0°C = 273 K. W: watt = 1 joule/s. m^2: square meter. Btu: British thermal unit = 1054.5 joules—the energy required to raise the temperature of 1 pound of water by 1°F.

TABLE 1.5 Typical UL Standards and ASTM Tests on Plastics

UL number	Standard
UL 94	Tests for Flammability of Plastic Materials for Parts in Devices and Applications
UL 746A	Polymeric Materials—Short Term Property Evaluations
UL 746B	Polymeric Materials—Long Term Property Evaluations
UL 746C	Polymeric Materials—Use in Electrical Equipment Evaluations
UL 746D	Polymeric Materials—Fabricated Parts
UL 746E	Polymeric Materials—Industrial Laminates, Filament Wound Tubing, Vulcanized Fibre, and Materials Used in Printed Wiring Board

ASTM number	Test method
Mechanical tests:	
D 256	Izod impact strength
D 638	Tensile properties
D 695	Compressive properties
D 785	Rockwell hardness
D 790	Flexural properties
D 953	Bearing strength
Chemical tests:	
D 543	Chemical resistance
D 570	Water absorption
Electrical tests:	
D 149	Dielectric strength
D 150	Dissipation factor and dielectric constant
D 257	Electrical resistance
D 495	Arc resistance
Thermal tests:	
D 648	Deflection temperature under load
D 696	Linear thermal expansion
D 746	Brittleness temperature
D 3417	Heats of fusion crystalline
D 3418	Transition temperatures
Flammability tests:	
D 635	Rate of burning
D 2843	Smoke density
D 2863	Oxygen index
E 662	Smoke emission

1.28 Glossary

ABS Acrylonitrile-butadiene-styrene terpolymers or blend.

acetic anhydride $H_3CCOOOCCH_3$.

acid, carboxylic A compound containing a carboxyl (COOH) group, such as formic acid (HCOOH).

addition polymerization A polymerization which consists of a sequential addition of vinyl monomers to growing active species.

Alcryn A TPE based on chlorinated polyethylene copolymers and EPDM.

alkyd A polyester resin produced by the condensation of phthalic anhydride, ethylene glycol, and an unsaturated vegetable oil.

aliphatic compound A linear organic compound, such as a paraffinic hydrocarbon derivative.

amide A compound produced by the condensation of an organic acid and an amine.

amine An organic derivative of ammonia.

aminoplastics UFs and MFs.

anhydride A compound produced by the dehydration of two carboxyl groups.

amorphous Noncrystalline.

anionic polymerization One where the initiating species is an anion (A:−).

Ardel A polyarylate resin.

aromatic compound A cyclic organic compound, such as benzene and its derivatives.

Astrel A polysulfone.

Baekeland, L. The inventor of PF plastics (Bakelite).

Bakelite A PF.

backbone The continuous polymer chain.

Bayblend An ABS-polysulfone blend.

Bayer, O. The inventor of polyurethanes.

bifunctional compound A compound, such as ethylene glycol, with two functional groups.

bisphenol A $HOC_6H_4C(CH_3)_2C_6H_4OH$.

block copolymer A copolymer consisting of sequences of repeating units of more than one polymer in a continuous chain (A_nB_n).

butyl rubber A copolymer of isobutylene and isoprene.

Cadon Blends and derivatives of SMA-AN terpolymer.

capping Reaction with end groups of a polymer.

caprolactam An amino carboxylic acid which tends to form a stable cyclic compound.

Carothers, W. H. An eminent polymer chemist who invented nylon and neoprene and made many contributions to modern polymer science.

catalyst A substance which changes the rate of a chemical reaction without undergoing permanent change in composition. Polymer initiators, which become part of the polymer chain, are erroneously called catalysts.

cationic polymerization One where the initiating species is a cation (M^+).

Celanex A PBT-PET blend.

Celcon A copolymer of formaldehyde and ethylene oxide.

CR-39 A polymer of a diallyl ester.

crystalline compound A solid characterized by a specific shape and arrangement of its atoms in a definite reproducible pattern.

cycloaliphatic hydrocarbon A saturated cyclic hydrocarbon.

Cycolac An ABS.

Dacron A polyester fiber (PET).

Dapon A polymer of diallyl phthalate.

Delrin A POM, a polymer of formaldehyde.

diadic polymer A polymer produced from the condensation of two reactants like nylon 6,6, also called AABB nylon.

Dickson, J. T. A coinventor of polyester fibers.

diol A dihydroxy compound.

drier An organic salt of a heavy metal which serves as a catalyst in the polymerization (drying) of oleoresinous coatings.

drying Polymerization of an unsaturated oil in the presence of oxygen.

Duprene The original trade name for neoprene.

Durel A polyarylate resin.

Dylark Blends and derivatives of SMA.

Elastopar A copolymer of butyl acrylate and α-chloroethyl vinyl ether.

EPDM A vulcanizable elastomeric copolymer of ethylene and propylene.

Farbenindustrie, I. G. A large German chemical corporation.

fibrous glass Glass filaments.

filament A continuous thread extruded from spinnerets.

Fox, D. W. The developer of commercial polycarbonate.

free radical An electron-deficient molecule such as $CH_3\bullet$, which is designated $R\bullet$.

glass transition temperature (T_g) The temperature at which a glassy polymer becomes flexible as the temperature is increased; T_g is sometimes called the second-order transition temperature.

glycol A dihydric compound, such as ethylene glycol.

Glyptal A polyester produced by the condensation of phthalic anhydride and glycerol.

graft copolymer A copolymer with chain extension of another polymer attached to various carbon atoms on the principal polymer chain.

Hay, A. The inventor of PPO.

hevea rubber Natural rubber (NR).

hexa Hexamethylenetetramine $(CH_2)_6N_4$.

Hill, Julian A Du Pont chemist who, in cooperation with Carothers, produced nylon 6,6 polymer.

hips High-impact polystyrene.

Hytrel A thermoplastic polyester elastomer.

injection molding A process in which a measured amount of molten polymer is forced into a closed split mold, where it solidifies by cooling and is injected when the mold is opened.

ionomer An ionic copolymer of ethylene and methacrylic acid (Surlyn).

Kalrez A copolymer of perfluoromethyl vinyl ether and TFE.

Kel-F A polychlorotrifluoroethylene (PCTFE).

Kienle, R. H. Assigned the acronym "alkyd" to polyesters produced from *al*cohol and *acids*.

Kipping, F. S. The inventor of silicones.

Koroseal A plasticized PVC.

Kralastic An ABS.

Kraton A block copolymer of polystyrene-polybutadiene and polystyrene.

Krazy Glue An adhesive based on α-cyanoacrylic esters.

Kynar A polyvinylidene fluoride, PVDF.

ladder polymer A polymer with a double backbone.

Lexan A polycarbonate (PC).

linear polymer A polymer with a continuous backbone, also called a straight chain.

linoleic acid $C_{17}H_{31}COOH$; an unsaturated high-molecular-weight acid present in linseed oil.

linolenic acid $C_{17}H_{29}COOH$; an unsaturated high-molecular-weight acid present in linseed oil.

Lucite A polymethyl methacrylate (PMMA).

Marbon An ABS.

melting point The temperature at which crystals and liquid are in equilibrium.

Merlon A polycarbonate (PC).

MF Melamine-formaldehyde resin.

Mindel An ABS-polysulfone blend.

molding powder A mixture of polymers and appropriate additives which can be molded by heating under pressure in a mold.

monadic polymer A polymer produced from one reactant, like nylon 6; also called AB type.

NBR Acrylonitrile-butadiene copolymer.

neoprene Poly-2-chloro-butadiene; $\leftarrow CH_2CCl{=}CHCH_2 \rightarrow$.

Noryl A PPO blend.

novolacs PF resins produced by the acid condensation of phenol and formaldehyde.

olefins Derivatives of ethylene.

oleoresinous paints Polymerizable coatings systems containing vegetable oils with unsaturated groups.

Ostromislensky, I. A Russian-American pioneer in polymer science.

Patrick, J. C. The inventor of America's first synthetic elastomers (Thiokol).

PAI Polyamideimide.

PBI Polybenzimidazole.

PBT Polybutylene terephthalate.

PEBA Polyether block amide engineering polymer.

Pebax A PEBA.

PEEK Polyether ether ketone.

PEI Polyetherimide.

pendant groups Functional groups on the polymer chain.

peracetic acid H_3COOOH.

peroxide A compound, such as benzoyl peroxide [$(CH_6H_5CO)_2O_2$], which contains more oxygen than its stable precursor, in this case benzoic acid (C_6H_5COOH).

PF Phenol-formaldehyde polymer.

PI Polyimide.

plant, pilot A small-scale production unit which is larger than laboratory scale and smaller than a commercial production unit.

plasticizer A high-boiling-point liquid used to flexibilize polymers, such as PVC.

Plioflex A copolymer of vinyl chloride and vinylidene chloride.

Plunkett, Roy The discoverer of Teflon.

PMMA Polymethyl methacrylate.

polyamic acid A linear prepolymer for polyimides.

polyamide A polymer produced by the condensation of a dicarboxylic acid and a diamine.

polyester A polymer produced by the condensation of a dicarboxylic organic acid and a dihydric alcohol (diol).

polyimide A heterocyclic polymer.

polyphosphazene $+N{=}P(OR)_2+$.

polyurea The reaction product of an organodiisocyanate and a diamine.

polyurethanes Polymers produced by the reaction of an organodiisocyanate and a diol.

POM Polyoxymethylene $[+CH_2O+]$.

prepolymer A low-molecular-weight compound that can be converted to a high-molecular-weight polymer.

primary functional group A group on a carbon atom which is attached to two hydrogen atoms.

reaction injection molding (RIM) The formation of solid polymer articles by the polymerization of the reactants in the mold.

resoles PF resins produced by the alkaline condensation of a phenol and a formaldehyde.

RIM Reaction injection molding.

Rohm, O. A long-time researcher and developer of acrylic polymers.

Rynite A polyester engineering polymer.

SAN Styrene-acrylonitrile copolymer.

Santoprene A TPE based on chlorinated polyethylene copolymers and EPDM.

SBR Styrene-butadiene elastomer.

Schlack, Paul The inventor of nylon 6.

Schnell, H. The developer of commercial polycarbonate.

Semon, W. A pioneer in the plasticization of PVC.

silicone $+OSiR_2+$ (siloxane).

Solprene A block copolymer of polystyrene and polybutadiene.

Sparks, William A coinventor of butyl rubber.

Styralloy A blend of SBR and PS.

Teflon Polytetrafluoroethylene.

Terylene A polyester fiber.

T_g Glass transition temperature.

Thermoplastic A linear polymer which may be reversibly softened by heat and solidified by cooling.

thermoset A three-dimensional insoluble polymer which cannot be melted or softened by heating.

Torlon A polyamideimide (PAI).

TPE Thermoplastic elastomer.

UF Urea-formaldehyde resin.

Ultem A polyether amide (PEI).

U polymer A polyarylate resin.

urea H_2NCONH_2.

urethanes Compounds produced by the reaction of organoisocyanates and alcohols. One of the simpler members of this family is phenylurethane ($C_6H_5NHCOOC_2H_5$).

Viton A copolymer of vinylidine fluoride and hexafluoropropylene.

Whinfield, J. R. A coinventor of polyester fibers.

1.29 References

Dubois, J. Harry: *Plastics History USA,* Cahners Publishing Company, Boston, 1972.

Epel, J. N., J. M. Margolis, S. Newman, and R. B. Seymour (eds.): *Engineering Plastics,* ASM International, Metals Park, Ohio, 1988.

Kaufman, M.: *The History of PVC,* Gordon and Breach, Science Publishers, Inc., New York, 1968.

Margolis, J. M.: *Engineering Thermoplastics,* Marcel Dekker, Inc., New York, 1985.

Martino, R. J.: *Modern Plastics* 65 (1):95 (1988).

Seymour, R. B.: *Applications of Polymers,* Plenum Publishing Corporation, New York, 1988.

Seymour, R. B.: *High Performance Polymers,* VNU Science Press, Utrecht, Holland, 1989.

Seymour, R. B. (ed.): *History of Polymer Science and Technology,* Marcel Dekker, Inc., New York, 1982.

Seymour, R. B.: *Plastics for Engineering Applications,* ASM International, Metals Park, Ohio, 1987.

Seymour, R. B., and C. E. Carraher: *Polymer Chemistry: An Introduction,* 2d ed., Marcel Dekker, Inc., New York, 1988.

Seymour, R. B., and C. E. Carraher: *Giant Molecules,* John Wiley & Sons, Inc., Interscience Publishers, New York, 1989.

Seymour, R. B., and R. D. Deanin (eds.): *Polymer Composites: Their Origin and Development,* VNU Science Press, Utrecht, Holland, 1987.

Seymour, R. B., and G. S. Kirshenbaum (eds.): *High Performance Polymers: Their Origin and Development,* Elsevier Science Publishing Company, Inc., New York, 1986.

Chapter

2

Phenolic, Furan, Urea, and Melamine Plastics

2.1 PF Polymers

Condensation products of phenol (carbolic acid) and formaldehyde (formalin) were made in the nineteenth century by several eminent organic chemists but were discarded as "goos, gunks, and messes" or waste products, because they were not crystalline compounds with characteristic melting points.

Leo Baekeland recognized that when trifunctional phenol was condensed with formaldehyde, it could produce a three-dimensional, infusible thermoset. Hence, he used care to prevent complete reaction when conducting the condensation of no more than two reaction sites of phenol on the alkaline side. These resole polymers (PF) could be thermoset by heating at higher temperatures or by the addition of strong acids, such as *p*-toluene sulfonic acid ($H_3CC_6H_4SO_3H$). A reaction at the third reaction site occurs at the higher temperature or in the presence of acids.

Baekeland also produced linear novolac PF condensation products by using an excess of phenol under acid conditions. The additional formaldehyde required for the production of a cross-linked plastic was added to the molding compound as hexa. The tendency for the PF resins to bubble in the mold was overcome by the use of pressure, as well as heat, as described in Baekeland's classic "heat and pressure" patent. The characteristic brittleness of PF was overcome by the addition of wood flour, i.e., finely divided fibrous wood made by attrition grinding on a buhr mill.

The term "compound" is used by the plastics and rubber technolo-

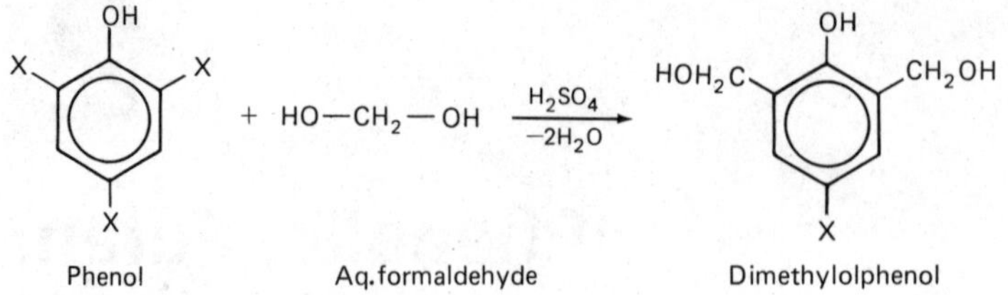

Phenol + Aq.formaldehyde $\xrightarrow[-2H_2O]{H_2SO_4}$ Dimethylolphenol

Figure 2.1

gist to describe a mixture of polymer and essential additives. PF novolac compounds are produced by adding a mold release agent such as zinc stearate, wood flour as a filler, pigment, and hexa as a source of formaldehyde to the PF prepolymer. The mixture is heated under pressure on a two-roll mill or in an extruder to induce additional polymerization, i.e., advance the resin to a higher-molecular-weight prepolymer which is then ground and used as a molding compound.

As shown in Fig. 2.1, phenol has three reactive hydrogen atoms (i.e., a functionality f of 3) on the benzene ring, in the 2, 6, and 4 positions, i.e., ortho, ortho, and para positions, which are, for convenience, designated by X. The accepted mechanism for the acid condensation involves a cation, produced by the protonation of formaldehyde (CH_2OH^+). It is believed that the alkaline condensation involves a condensation of formaldehyde with the sodium salt of phenol, sodium phenoxide (C_6H_5ONa). However, for simplicity, a less sophisticated mechanism will be used in this chapter.

Aqueous formaldehyde may exist as a methylene diol ($HOCH_2OH$), which can react with the active hydrogen atóms on the benzene molecule to produce methylol groups ($—CH_2OH$), as shown in Fig. 2.1.

The dimethylolphenol may react with similar molecules to produce a linear prepolymer, if the phenol is in excess of that required to permit condensation with the hydrogen atom in the number 4, or para, position, as shown in Fig. 2.2.

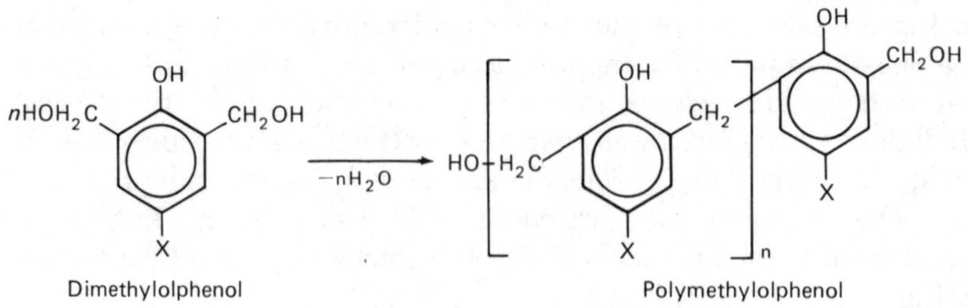

Dimethylolphenol $\xrightarrow[-nH_2O]{}$ Polymethylolphenol

Figure 2.2

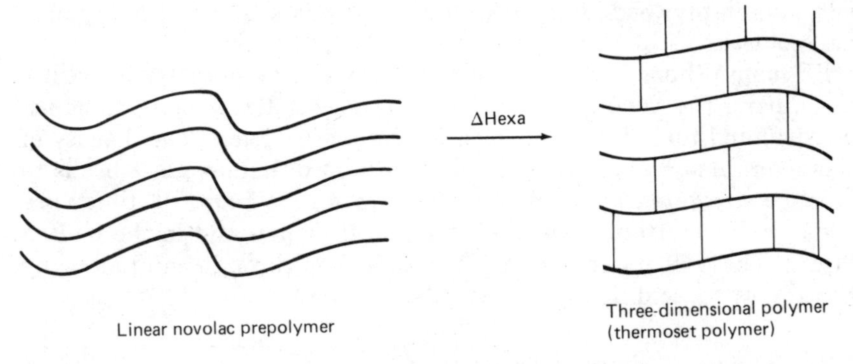

Figure 2.3

This linear prepolymer is called an A-stage novolac resin. This prepolymer forms a B-stage novolac (or resitol) resin when blended and heated at moderate temperatures with hexa. The phenolic B-stage resin is blended with unreactive additives, such as wood flour, dyes or pigments, and mold release agents, and heated to produce a cross-linked (three-dimensional) plastic (or resist) C-stage resin in the mold. The cross-linking step is shown in the sketch in Fig. 2.3.

If this condensation takes place with anhydrous formaldehyde, dimethylene ether groups $\text{+}CH_2OCH_2\text{+}$ may be present in the polymer backbone along with the methylene (CH_2) groups.

The resole phenolic polymers are single-stage resins which are produced by the controlled alkaline condensation of phenol and formaldehyde. Since adequate amounts of formaldehyde are present, the condensation proceeds through a linear prepolymer to a cross-linked thermoset, or network, polymer. The rate of the cross-linking reaction can be controlled by reducing the temperature. Thus, liquid resole resins can be stored for several weeks at temperatures below 0°C. The rate of this condensation can be enhanced by the addition of strong mineral acids or by heating.

2.2 Processing of Phenolic Polymers

Resole resins may be used as binders for fiberglass, fiberboard, brake and clutch bindings, plywood, lacquers, molds for metal castings, vulcanized fiber, aggregates in grinding wheels, and foundry sand cores, and as molding powders. These resins are dark and may be heat-hardened at 140 to 180°C.

Over 1,300,000 t of PF resins was consumed in the United States in 1987. Almost 50 percent of this production was used as an adhesive in plywood, and only 7.5 percent was used for molding phenolic plastics.

PF foam (Phenofoam), produced by blowing the prepolymer with a gas during the curing process, has the lowest flame spread, smoke density, and toxic fume emission of any plastic foam. The density of this foam may be controlled by the addition of hollow glass beads to produce a low-density product called a syntactic foam. PF fibers are being produced from novolac PF (Kynol) in Japan and in the United States. These fibers are cured, after extrusion, through spinnerets, by heating in an acidified formaldehyde solution.

2.3 Molding of Phenolic Polymers

It has been the custom to produce various PF articles by compression molding in a heated two-piece mold which has a mold cavity shaped like the finished article. In this pioneer molding process, a measured amount of the PF compound is heated in the lower half of the mold and the heated compound is forced into the boundaries of the mold cavity by pressure exerted by the upper half of the heated mold. The top section of the mold is called the force (male plug) and the bottom section is called the cavity (female cavity). The molding press consists of two flat surfaces (platens), and the distance between the platens is designated by the term "daylight opening."

When full pressure is directed to the molding compound, the mold is called a positive mold. In semipositive molds, some pressure is exerted on the upper surface of the mold (land) and some excess plastic (flash) is allowed to escape over the land surface. Larger amounts of flash are allowed to escape in less costly flash molds.

The rate of cure (cross-linking) in the molding process is directly proportional to the molding temperature. The heat may be supplied by electric resistance or hot oil. It is customary to use a temperature of 149°C (300°F) under a pressure of 21 megapascals (MPa) [3000 pounds per square inch (lb/in^2)] and a molding time of 1.5 minutes (min). After curing, the PF molded article is removed by opening the mold.

In a modified molding procedure called transfer molding, the preform (compressed briquet) of the PF compound is preheated by high-frequency dielectric heating and forced by a ram into a pot and then through runners (channels) and gates (narrow openings) into heated mold cavities. The molded articles are removed from the mold and are separated from the runners and sprue. (The sprue is a cured slug of polymer which is formed when the resin is cured in the vertical entrance to the cavity.)

The molds used in transfer molding are more expensive than those used in compression molding, and the sprues and runners, which are thermoset, must be discarded. However, multiple mold cavities can be used in the transfer-molding process, and the curing cycle is much shorter than that in conventional compression molding.

The molding cycles may be reduced further by injection molding. In the more modern direct screw transfer (DST) process, a reciprocating screw (preplasticizer) plasticates (softens) the molding compound and stops turning when the desired amount of PF molding compound reaches the water-cooled nozzle end of the barrel. The screw is then moved forward so that it forces the molten PF through the sprue into the runners and gates and finally into the mold cavity.

The molding cycle, i.e., the time interval between die closings, is preset in accordance with the time required for the thermoset molding to become rigid. In contrast, this cycle for a thermoplastic is controlled by the time required for the plastic part to cool so that it is rigid. As is the case with multiple cavity transfer molds, the ejected part is attached to runners and the sprue bushing, which must be detached from the molded parts. Injection molding is the most capital-intensive, the least labor-intensive, and the most widely used process for molding PF and other polymers.

Waste associated with discarded thermoset sprues may be eliminated by using modified injection-molding techniques. Scrap loss is also eliminated by the use of runnerless injection-compression (RIC) molding in which the PF is injected into a partially opened mold which is closed during the compression-molding processing. Sprues, etc., from thermoplastics may be ground and remolded.

2.4 Wood-Based Composites

Insulating board, hardboard, paperboard, particleboard, and plywood are composites of wood or paper and PF resole adhesives. Plywood, which is the principal PF product, is produced by using the liquid resole adhesive as a glue to bond an odd number of thin layers of wood (veneer) at right angles to each other. This isotropic composite is widely used in place of thicker wooden boards.

Laminates consisting of parallel layers of wood joined by PF adhesives are used in place of wood, especially when long and/or curved members must be fabricated. Structural glued-laminated tubes (glulam) consisting of multiple layers of wood adhered by PF adhesives are used for roof-supporting structures. Particleboard is produced by molding a mixture of PF resin and wood chips.

2.5 Compounding of PF

Resins produced by the condensation of formaldehyde and resorcinol (1,3-dihydroxybenzene) cure more rapidly than PF. Cashew nut oil, which is a meta-substituted (substituent on carbon 3) phenol, also reacts more readily with formaldehyde than phenol. Furfural (C_4H_3OCHO), which is obtained from corncobs, may be used in place of formaldehyde to produce a molding powder with better melt flow than PF.

In addition to wood-flour filler, which was used by Baekeland to produce Bakelite, asbestos, cotton fiber, and fiberglass have been used as reinforcements for PF. Mineral-filled PF with a heat deflection temperature of 215°C (390°F) is also available commercially. PF is also used as a matrix for reinforced plastics.

2.6 Properties of PF

As shown by the data in Table 2.1, PF has good resistance to organic solvents, aqueous salt solutions, and nonoxidizing acids, such as hydrochloric acid. Because of the presence of an acidic hydrogen atom on the hydroxyl substituent, PF is not resistant to strong alkalies and hence will swell when immersed in hot aqueous alkaline solutions.

2.7 Applications of PF

Phenolic molding powder (a granular material) is usually a novolac resin (two-step). The general-purpose PF is wood-flour-filled, but clay, mica, and cellulose may be used as fillers, and graphite fibers and glass fibers may be used as reinforcements. Molded PF composites have excellent creep resistance and very good dimensional tolerance.

PF is the workhorse of thermosets. In spite of its dark color, it is economical, and it is characterized by excellent mechanical, thermal, and electrical properties. Fiberglass-reinforced PFs have superior dimensional stability, and their resistance to heat and impact are superior to general-purpose, wood-flour-filled PF. Fiberglass-filled PFs retain their properties over long periods of time at temperatures as high as 232°C (450°F). Special PF resins, such as phenylsilanes, can be used at temperatures as high as 538°C (1000°F) for short periods of time.

Ablative (silica-fiber-filled) PF, used in aerospace applications, can withstand a temperature of 2815°C (6000°F) for a few minutes and a temperature of 13,930°C (25,000°F) for a few seconds.

Over 2,700,000 t of phenolic resins was produced in the United States in 1987. The principal end uses of PF are as plywood (47.5%), insulation (16%), wood composites (9%), and molded plastics (7.5%).

Because of its good electrical properties, molded PF is used in many electrical applications, such as toasters, pot handles, wiring devices, and circuit breakers. Because of its good resistance to heat, PF is used

TABLE 2.1 Properties of Typical Phenolic Plastics (PF)

Property, units*	Wood-flour filled	40% Graphite-fiber-reinforced	40% Fiberglass-reinforced
Melting point T_m, °F	—	—	—
Glass transition temp. T_g, °F	—	—	—
Processing temp., °F	300–375	300–350	300–375
Molding pressure, 10^3 lb/in^2	2–20	2–5	1–20
Mold shrinkage, 10^{-3} in/in	4–9	0.5	1.4
Heat deflection temp. under flexural load of 264 lb/in^2, °F	340	480	475
Maximum resistance to continuous heat, °F	300	375	375
Coefficient of linear expansion, 10^{-6} in/(in · °F)	30	20	10
Compressive strength, 10^3 lb/in^2	30	30	20
Impact strength, Izod, ft · lb/in of notch	0.4	10	12
Tensile strength, 10^3 lb/in^2	—	—	—
Flexural strength, 10^3 lb/in^2	110	36	40
Percent elongation	0.6	0.4	0.2
Tensile modulus, 10^3 lb/in^2	1000	—	2500
Flexural modulus, 10^3 lb/in^2	1100	415	2000
Rockwell hardness	M110	M115	E75
Specific gravity	1.4	1.45	1.8
Percent water absorption	0.6	0.2	0.5
Dielectric constant	6	8	5
Dielectric strength V/mil	325	—	375
Resistance to chemicals at 75°F:†			
Nonoxidizing acids (20% H_2SO_4)	S	S	S
Oxidizing acids (10% HNO_3)	Q	Q	Q
Aqueous salt solutions (NaCl)	S	S	S
Polar solvents (C_2H_5OH)	S	S	S
Nonpolar solvents (C_6H_6)	S	S	S
Water	S	S	S
Aqueous alkaline solutions (NaOH)	U	U	U

*lb/(in^2 · 0.145) = KPa (kilopascals); ft · lb/(in of notch · 0.0187) = cm · N/cm of notch.
†S = satisfactory; Q = questionable; U = unsatisfactory.

as a component of heated appliances, disc brakes, and under-the-hood applications. The graphite- and fiberglass-reinforced PF are used for transmission stators, automotive water pumps, brake pistons, wiring devices, and down-hole oil well applications.

2.8 Phenolic Laminates

Plywood, which is the most widely used PF laminate, is produced by adhering relatively thin layers or plies of wood (veneers) so that the grains of adjacent layers are perpendicular to each other, i.e., at 90°. It is customary to cut the veneer from steam-heated logs by the use of a rotary lathe. The veneer is then dried and coated with a resole PF resin, and an odd number of layers are then heated under pressure in a platen press.

Placing the veneers at right angles provides isotropicity (similar properties in all directions) and permits the use of thinner layers than boards, which are anisotropic. Laminated wood with parallel-grain construction is also available. This anisotropic laminate is used as an engineered structure for large members, such as arches and beams.

Laminates of three or more layers of sawed lumber (glulam) have been used for sporting goods and straight or arched structural beams, such as roof-supporting structures.

2.9 Toxicity of PF

While both phenol and formaldehyde are toxic, there is no free phenol or formaldehyde in properly cured PF. Hence, these cured PF resins are used as coatings in beer and soft drink cans. No detrimental effects of PF polymers have been reported in the literature. PF has moderate resistance to flames, and this resistance can be enhanced by the addition of flame-retardant fillers, such as alumina trihydrate (ATH).

2.10 Furan Plastics

Those interested in chemurgy, renewable resources, and biomass sources of polymers have championed the use of furan plastics, which are derived from pentosans in agricultural wastes, such as corncobs, oat hulls, or wood. These pentosans are readily hydrolyzed by hot aqueous acids, such as hydrochloric acid, to produce furfural, which can be used in place of formaldehyde in the production of phenolic plastics.

Furfural may be hydrogenated to produce furfuryl alcohol ($C_4H_3OCH_2OH$). A black, infusible polymer is produced when furfural is heated at 204°C (400°F) in an inert atmosphere. Furfural condenses, under acid or alkaline conditions, with phenol, bisphenol A, and acetone (H_3CCOCH_3). These furan resins have been used as coatings, adhesives for wood particles, and silica-filled composites.

The most widely used furan resins are based on acid-polymerized furfuryl alcohol. These dark polymers have been used as metal-casting cores and molds, corrosion-resistant mortars, and polymer concrete. The mortars (Alkor) have been used to join brick and tile in chemical processing equipment and floors. Furan resins have also

TABLE 2.2 Properties of a Typical Carbon-Filled Furan Resin

Property, units*	
Melting point T_m, °F	—
Glass transition temp. T_g, °F	—
Processing temp., °F	—
Molding pressure, 10^3 lb/in^2	—
Mold shrinkage, 10^{-3} in/in	—
Heat deflection temp. under flexural load of 264 lb/in^2, °F	175
Maximum resistance to continuous heat, °F	210
Coefficient of linear expansion, 10^{-6} in/(in · °F)	40
Compressive strength, 10^3 lb/in^2	15
Impact strength, Izod, ft · lb/in of notch	0.5
Tensile strength, 10^3 lb/in^2	6
Flexural strength, 10^3 lb/in^2	7
Percent elongation	1.5
Tensile modulus, 10^3 lb/in^2	—
Flexural modulus, 10^3 lb/in^2	—
Rockwell hardness	R110
Specific gravity	1.7
Percent water absorption	—
Dielectric constant	—
Dielectric strength V/mil	—
Resistance to chemicals at 75°F:†	
Nonoxidizing acids (20% H_2SO_4)	S
Oxidizing acids (10% HNO_3)	U
Aqueous salt solutions (NaCl)	S
Polar solvents (C_2H_5OH)	S
Nonpolar solvents (C_6H_6)	S
Water	S
Aqueous alkaline solutions (NaOH)	S

*lb/(in^2 · 0.145) = KPa (kilopascals); ft · lb/(in of notch · 0.0187) = cm · N/cm of notch.
†S = satisfactory; Q = questionable; U = unsatisfactory.

been used to stabilize loose sands in petroleum wells. Over 200,000 t of furfural was produced worldwide in 1987.

The properties of a typical carbon-filled furan resin are shown in Table 2.2.

2.11 Production of Amino Resins, Adhesives, and Laminates

The production procedures for UF and MF resins are similar and are related to that of PF. Hence, UF and MF are often classified as amino resins. Both are colorless resins, but UF accounted for 80 percent of the 725,000 t of amino resins produced in the United States in 1987.

Most of the urea (H_2NCONH_2) produced is used as fertilizer and other inexpensive applications, and it is a low-cost material; this is reflected in the economical production of UF. Since melamine is produced by the trimerization of cyanamide, which is obtained by the dehydration of urea, it is also relatively inexpensive but more costly than urea.

Both UF and MF are produced by the acid condensation of the methylol derivatives (—$NHCH_2OH$) of urea and melamine, which are obtained by the acid or alkaline condensation of these reactants with formaldehyde. Since the methylol prepolymers of UF and MF are soluble in water, they are readily used as polymerizable coatings, adhesives, and molding compounds. Over 450,000 t of amino resins was used for bonding wood, 50,000 t for plywood and other laminates, and 30,000 t as molding resins in the United States in 1987.

Amino resin adhesives may be cured at room temperature by the addition of ammonium chloride (NH_4Cl) or by electron beam (EB) radiation. The water resistance of MF is much better than that of UF. It is customary to use PF resins for the interior layers of laminates and to place the lighter-colored UF or MF resins on the laminated surface. These laminates are cured between steam-heated plates under a pressure of 0.7 MPa (100 lb/in^2) at 149°C (300°F).

2.12 Amino Resin Molding Compounds

Granular amino resin molding compounds are produced by mixing the prepolymers with α-cellulose pulp, drying, grinding, mixing with pigments, etc., and compacting under pressure. α-Cellulose is a high-molecular-weight product which is insoluble in 17.5% caustic solution (NaOH). As was described for the transfer molding of PF, it is customary to preheat a preform of the molding powder before placing it in the mold cavity. MF has excellent electrical properties, i.e., a dielectric constant at 10^6 hertz (Hz) = 7.0, dissipation factor at 10^6 Hz = 0.04, and an arc resistance of 180 seconds (s). Other properties of typical α-cellulose-filled amino resins are shown in Table 2.3.

2.13 Toxicity

Urea, melamine, and their stable polymers (UF, MF) are nontoxic. Molded MF dinnerware does not release formaldehyde, but some amino resins, such as UF foamed-in-place insulation, may release formaldehyde, which, even in low concentrations, is an eye, nose, and throat irritant. Exposure to large quantities of formaldehyde over long periods of time has produced cancer in laboratory animals. Hence, exposure to formaldehyde fumes should be avoided.

Formaldehyde, hydrogen cyanide (HCN), and nitrogen oxides may

TABLE 2.3 Properties of Typical α-Cellulose-Filled Urea Plastics (UF)

Property, units*	
Melting point T_m, °F	—
Glass transition temp. T_g, °F	—
Processing temp., °F	325
Molding pressure, 10^3lb/in^2	15
Mold shrinkage, 10^{-3} in/in	8
Heat deflection temp. under flexural load of 264 lb/in^2, °F	275
Maximum resistance to continuous heat, °F	250
Coefficient of linear expansion, 10^{-6} in/(in · °F)	15
Compressive strength, 10^3 lb/in^2	35
Impact strength, Izod, ft · lb/in of notch	0.3
Tensile strength, 10^3 lb/in^2	10
Flexural strength, 10^3 lb/in^2	10
Percent elongation	0.5
Tensile modulus, 10^3 lb/in^2	1000
Flexural modulus, 10^3 lb/in^2	1500
Rockwell hardness	M115
Specific gravity	1.15
Percent water absorption	0.6
Dielectric constant	6
Dielectric strength V/mil	350
Resistance to chemicals at 75°F:†	
Nonoxidizing acids (20% H_2SO_4)	Q
Oxidizing acids (10% HNO_3)	U
Aqueous salt solutions (NaCl)	S
Polar solvents (C_2H_5OH)	S
Nonpolar solvents (C_6H_6)	Q
Water	S
Aqueous alkaline solutions (NaOH)	Q

*lb/(in^2 · 0.145) = KPa (kilopascals); ft · lb/(in of notch · 0.0187) = cm · N/cm of notch.
†S = satisfactory; Q = questionable; U = unsatisfactory.

be produced during the combustion of amino resins. However, these resins are classified as self-extinguishing and nonburning.

2.14 Glossary

acid, nonoxidizing Hydrochloride, phosphoric, and dilute sulfuric acids.

Alkor A carbon-filled furan cement.

alpha (α)-cellulose A high-molecular-weight cellulose filler.

amino resins Urea or melamine-formaldehyde resins.

anhydrous Water-free.

anisotropic Having properties that vary with direction, i.e., nonuniform.

A-stage resin A low-molecular-weight linear prepolymer.

Baekeland, L. The inventor of PF plastics (Bakelite).

Bakelite A PF.

B-stage resin A medium-molecular-weight linear prepolymer.

carbolic acid Phenol.

cashew nut oil A meta-substituted phenol (Cardanol) obtained from the shells of cashew nuts (Anacardium occidentale). This oil is a vesicant and causes blistering if it is in contact with the skin.

compound A mixture of resin and other additives used for molding.

cross-linked polymer A three-dimensional polymer.

cyanamide H_2NCN.

DST Direct screw transfer molding.

formaldehyde H_2CO.

formalin An aqueous solution of formaldehyde, usually 37%.

furan plastics Those based on furfural or furfuryl alcohol.

furfural An aldehyde of the heterocyclic compound furan, which is obtained by the acid distillation of pentosans found in corncobs.

glulam A structural glued-laminated timber.

heat deflection temperature The temperature, under flexural load (DTUL), at which a specified deflection of a plastic bar occurs (ASTMD648-72/78).

hexa Hexamethylenetetramine.

hexamethylenetetramine $(CH_2)_6N_4$: a condensation product of ammonia and formaldehyde.

isotropic Having properties that are similar in all directions, i.e., uniform properties.

KPa Kilopascals (pressure).

Kynol A PF fiber.

laminate A composite consisting of layers of resins and nonresinous sheets.

linear Having a continuous chain of covalently bonded atoms.

melamine A cyclic trimer of cyanamide $C_3N_3(NH_2)_3$.

methylol group —CH_2OH—.

MF Melamine-formaldehyde resin.

molding, compression A process in which a molding powder is heated under pressure in a two-piece mold.

molding, injection A process in which a measured amount of molten polymer is forced into the cavity of a closed two-piece mold and then ejected as a cooled solid after the mold is opened.

molding, transfer A process in which a preheated preform is forced through runners and gates into the mold cavity.

N Newton (force).

nonpolar solvent An organic water-immiscible liquid like gasoline or benzene.

novolac A phenolic resin produced by the condensation of phenol and formaldehyde under acid conditions.

pentosans Complex carbohydrates (hemicelluloses) which are polymers of five-carbon sugars (pentoses), such as xylose.

PF Phenol-formaldehyde polymer.

Phenofoam A PF foam.

phenol Hydroxybenzene: C_6H_5OH.

polar solvent An organic liquid that is usually water-soluble.

prepolymer A partially cured polymer.

PSI Pounds per square inch; lb/in^2.

resole A phenolic resin produced by the condensation of phenol and formaldehyde under alkaline conditions.

resorcinol 1,3-Dihydroxybenzene.

RIC Runnerless injection-compression molding.

runners Narrow channels through which molten polymer is conveyed from the sprue to the mold cavity.

spinnerets Small holes through which a resin is forced to produce filaments.

sprue A channel in the molding press which connects the nozzle to the runner in the mold cavity. Term also used for resin which has solidified in sprue cavity.

syntactic foam One containing hollow glass beads.

wood flour A fibrous filler obtained by attrition grinding of wood.

UF Urea-formaldehyde resin.

urea H_2NCONH_2.

zinc stearate A mold release additive.

2.15 References

Blythe, A. R.: *Electrical Properties of Polymers,* Cambridge University Press, Cambridge, U.K., 1974.

Dunlap, A. P., and F. N. Peters: *The Furans,* ACS Monograph Series 119, Reinhold Publishing Corporation, New York, 1953.

Frados, J. (ed.): *Plastic Engineering Handbook,* Van Nostrand Reinhold Company, Inc., New York, 1976.

Gandini, A.: "Furan Resins," in J. I. Kroschwitz (ed.), *Encyclopedia of Polymer Science and Engineering,* 2d ed., vol. 7, John Wiley & Sons, Inc., Interscience Publishers, New York, 1987.

Maloney, T. M.: *Modern Particleboard and Dry Process Fiber Board Manufacturer,* Miller Freeman Publications, Inc., San Francisco, 1977.

Megson, N. J. L.: *Phenolic Resin Chemistry,* Academic Press, Inc., New York, 1958.
Updegraff, I. H.: "Amino Resins," in J. I. Kroschwitz (ed.), *Encyclopedia of Polymer Science and Engineering*, 2d ed., vol. 1, John Wiley & Sons, Inc., Interscience Publishers, New York, 1985.
Vale, C. P., and W. G. K. Taylor: *Amino Resins,* ILIFFE Books, Ltd., London, 1960.
Williams, L. L., I. H. Updegraff, and J.C. Petropoulos: chap. 45, in R. W. Tess and G. W. Poehlein (eds.), *Applied Polymer Science,* 2d ed., ACS Symposium Series 285, Washington, D.C., 1985.

Chapter

3

Thermoset Polyesters

3.1 Introduction

Cellulose nitrate (Celluloid), cellulose acetate, and the acrylates are thermoplastic polyesters. Glyptal, which is produced by the condensation of difunctional phthalic anhydride and trifunctional glycerol, was the first commercial thermosetting polyester, but its use was restricted primarily to protective coatings. The unsaturated polyesters are also polyesters, but since they are used primarily as composites and thermoplastics, they will be discussed in Chaps. 4 and 12, respectively.

3.2 Design Considerations

In contrast to metals, which have a long history of use and about whose properties there is much information, most engineering plastics and their composites are relatively new, and too little effort has been directed toward the development of information essential for the design of various products from polymers.

The properties of steel, which may be considered, in general, to be an elastic material, are governed by Hooke's law, in which the stress σ is proportional to the strain ϵ. The proportionality constant is the elastic modulus E. According to Hooke's law, the stress is independent of the rate of strain $d\epsilon/dt$, i.e, the stress is independent of the testing time.

In contrast, polymers are viscoelastic and must be considered as having the properties of both elastic and viscous materials. Hence, they are also governed by Newton's law, in which the stress is proportional to the rate of strain $d\epsilon/dt$, and the proportionality constant is the viscosity η. Thus, the rate of testing is a variable, and the stress-strain relations are much more complex than those in metals.

The modulus for viscoelastic materials σ/ϵ is a function of the rate of strain $d\epsilon/dt$ and one must be concerned with creep, stress relaxation, stress-strain rates, and vibrational loading. Creep is the result of an increase of strain over a long period of time. The stress diminishes, under constant strain, as a result of stress relaxation over a period of time. Vibrational effects are related to rapid cyclical reversible strain.

3.3 Reinforcement Concepts

The properties of a fiber-reinforced plastic composite are a function of the fiber properties, the concentration of the reinforcement, the alignment and the interfacial bonding of the resin matrix, and the fiber surface. The law of mixtures can be applied to the simplest model, i.e., one consisting of continuous parallel fibers embedded in the resin matrix. Accordingly, as shown by Eq. (3.1), the modulus of the composite, E_c, is equal to the modulus of the fiber, E_f, and of the matrix, E_m, multiplied by their partial volume V.

$$E_c = E_f V_f + E_m V_m \quad \text{or} \quad E_c = E_f V_f + E_m(1 - V_f) \tag{3.1}$$

Also, according to the law of mixtures, the ratio of the load carried by the reinforcing fibers, L_f, to that by the matrix, L_m, is equal to the ratio of their moduli, as shown in Eq. (3.2).

$$\frac{L_f}{L_m} = \frac{E_f V_f}{E_m(1 - V_f)} \tag{3.2}$$

When a large load is applied to a fiber-reinforced resin composite, some of the weaker fibers may fracture and, according to the uncoupled model, will not contribute to the reinforcement. However, the load can be transferred from the resin matrix to the fiber ends, and hence the uncoupled model is replaced by a coupled one in which one is concerned with the critical fiber length l_c.

In the Kelley and Tyson theory for the longitudinal stress buildup of the reinforcing fiber, it is assumed that the constant value of stress at the resin-fiber interface is limited by friction or plastic flow of the matrix, which occurs at an optimum shear stress σ, as shown in Eq. (3.3).

$$l_c = \frac{E_{fu} d_f}{2\sigma} \tag{3.3}$$

The critical fiber length l_c is directly proportional to the ultimate breaking strength of the fiber, E_{fu}, and the diameter of the fiber, d_f, and inversely proportional to twice the optimum shear stress σ.

If the fibers are all aligned, as in wood, the modulus measured on a weight basis in the fiber direction is equal to that of steel and, of course, is much greater than that of steel when measured on a volume basis. However, the modulus measured in other than the fiber direction may be as low as 3 percent of the maximum value. Of course, this difficulty is reduced by cross-lamination of composite layers.

3.4 Coupling Agents

In the concepts cited in Sec. 3.3, it is assumed that there is optimum adhesion at the interface between the resin matrix (continuous phase) and the fiber surface (stationary phase). Actually, the adhesion between glass and polyester resins is poor, and the glass fibers must be treated with coupling agents to improve the interfacial adhesion. The pioneer coupling agent (linking agent) was methacrylatochromic chloride (Volan). This additive has been supplemented by organosilanes, organotitanates, and organozirconates. These coupling agents contain functional groups, one of which is attracted to the filler or reinforcing fiber surface and the other to the resin.

3.5 Reinforcements

Glass filaments are produced by melting a mixture of silica, limestone, boric acid, and other reactants, such as fluorspar, at a temperature of 1260°C (2300°F) and forcing the molten product through small holes (bushings). The hot filaments are gathered together and cooled by a water spray. These multiple amorphous glass filaments are gathered together into a bundle, called a strand, which is wound up on a coil. Short fibers (staple) are produced by passing a stream of air across the filaments as they emerge from the bushings.

A high-alkali A-glass and a low-alkali E-glass are used as reinforcements for polymer composites. A more chemical-resistant glass, sodium borosilicate (C-glass), and a higher-tensile-strength glass, S-glass, are also available. E-glass is a calcium aluminosilicate, and S-glass is a magnesium aluminosilicate. Fiberglass is available as a collection of parallel filaments (roving), chopped strands, mat, and woven fabric.

Thomas Edison produced carbon (graphite) filaments for his original incandescent lamps by the carbonization of bamboo in the absence of air. Graphite filaments are now produced by the pyrolysis of polyacrylonitrile (PAN) filaments. High-tensile-strength graphite filaments are obtained by holding the PAN filaments under tension before exposing them to an oxidizing atmosphere. These fibers are then carbonized at temperatures on the order of 3000°C (5432°F) to obtain filaments with a high tensile strength.

TABLE 3.1 Properties of Typical Reinforcing Fibers

Property, units*	E-glass	S-glass	Graphite (Thornel 300)	Aramid (Kevlar 49)	Boron
Fiber diameter, mm	7.0	10.0	7.0	12.0	140
Tensile strength, 10^3 lb/in^2	500	700	500	400	515
Tensile modulus, 10^3 lb/in^2	10,400	12,300	33,500	19,000	56,000
Coefficient of expansion 10^{-6} in/(in · °F)	2.5	2.6	−1.0	− 1.2	2.6

*1 in = 0.254 mm; lb/in^2 × 0.0069 = MPa.

Kevlar 49 is a high-modulus aramid (aromatic nylon) fiber and is the most widely used reinforcing aramid fiber. This poly-*p*-phenylenephthalamide (PPP-T) is produced by the condensation of *p*-phenylenediamine and terephthaloyl chloride. The filaments of this liquid crystalline polymer are produced by forcing the molten aramid through spinnerets in a process called melt spinning. Kevlar 29 has a lower modulus and Kevlar 149 has a higher modulus than Kevlar 49. In some instances, a blend of graphite and aramid fibers, called hybrid fibers, is used to improve compression strength.

Boron composite filaments, produced by the chemical vapor deposition (CVD) of boron from a mixture of hydrogen and boron trichloride (BCl_3) onto a tungsten or graphite filament, have also been used as sophisticated reinforcements for polymeric composites. Silicon carbide and boron carbide have been investigated, but these have not been used for the production of commercial polymeric composites.

The properties of typical reinforcing fibers are shown in Table 3.1.

3.6 Diallyl Phthalate

Polydiallyl phthalate (DAP, which was one of the first commercial cast thermosetting polyesters, continues to be used today. Monoallyl esters can be polymerized to produce low-molecular-weight polymers (oligomers), but the diallyl esters produce higher-molecular-weight polymers. Polyesters with useful physical properties are obtained when low-molecular-weight polymers, such as DAP, are cross-linked.

In contrast to PF, UF, and MF, which are produced by condensation, or step-reaction, polymerization, DAP and other unsaturated monomers polymerize via addition, or chain, polymerization. It is customary to use an organic peroxide initiator (erroneously called a catalyst) and to precipitate the polymer by the addition of methanol after a viscous syrup has been produced.

As shown in Eq. (3.4), the first step in the polymerization of DAP and other unsaturated monomers, such as vinyl compounds, is the addition of the free radical (R ·), produced by the cleavage of the peroxide linkage $-(O-O)-$, to the DAP. For simplicity, the monomer will be shown as $H_2C{=}CHX$.

$$\underset{\text{Free radical}}{R\cdot} + \underset{\text{Monomer}}{H_2C{=}CHX} \rightarrow \underset{\text{New free radical}}{RCH_2CHX\cdot} \tag{3.4}$$

As shown by Eq. (3.5), the free radical, formed by the addition of R · to the monomer, adds rapidly to other monomers to produce a macroradical in sequential propagation steps.

$$RCH_2CHX\cdot + nH_2C{=}CHX \rightarrow \underset{\text{Macroradical}}{R(CH_2CHX)_nCH_2CHX\cdot} \tag{3.5}$$

As shown by Eqs. (3.6) and (3.7), the propagation reaction may be terminated by the coupling of two macroradicals to produce a high-molecular-weight stable polymer (dead polymer) or by reaction with an additive containing a labile atom (chain-transfer agent), such as hydrogen. This additive may donate the hydrogen atom to the macroradical to form a dead polymer and a new free radical in a chain-transfer reaction as shown in Eq. (3.7).

$$2R(CH_2CHX)nCH_2CHX\cdot \rightarrow \underset{\text{Dead polymer}}{R(CH_2CHX)_{n+1}(XHCCH_2R)_{n+1}} \tag{3.6}$$

$$R(CH_2CHX)_nCH_2CHX\cdot + \underset{\text{Chain transfer agent}}{HA} \rightarrow \underset{\text{Chain-transfer agent}}{R(CH_2CHX)_nCH_2CH_2X} + \underset{\text{New free radical}}{A\cdot} \tag{3.7}$$

When two active double bonds are present in a monomer, such as in DAP, the propagation may continue to produce a cross-linked polymer. Diallyl isophthalate (meta, or 1,3, isomer) polymerizes faster than the conventional ortho (1,2) ester.

In the preparation of molding powders from DAP, the precipitated polymer is dried and mixed with DAP, additional initiator, and other additives, such as fillers, mold release agents, and pigments. The molding compound is then heated in a compression transfer or injection mold similar to those described for PF in Chap. 2.

The properties of a fiberglass-reinforced polymer of DAP are shown in Table 3.2.

TABLE 3.2 Properties of a Typical Fiberglass-Reinforced DAP Polymer

Property, units*	
Melting point T_m, °F	—
Glass transition temp. T_g, °F	—
Processing temp., °F	290
Molding pressure, 10^3lb/in^2	20
Mold shrinkage, 10^{-3} in/in	5
Heat deflection temp. under flexural load of 264 lb/in^2, °F	400
Maximum resistance to continuous heat, °F	300
Coefficient of linear expansion, 10^{-6} in/(in · °F)	20
Compressive strength, 10^3 lb/in^2	30
Izod impact strength, ft · lb/in of notch	4
Tensile strength, 10^3 lb/in^2	9
Flexural strength, 10^3 lb/in^2	15
Percent elongation	4
Tensile modulus, 10^3 lb/in^2	15,000
Flexural modulus, 10^3 lb/in^2	18,000
Rockwell hardness	E85
Specific gravity	1.8
Percent water absorption	0.2
Dielectric constant	4
Dielectric strength, V/mil	400
Resistance to chemicals at 75°F:†	
Nonoxidizing acids (20% H_2SO_4)	S
Oxidizing acids (10% HNO_3)	Q
Aqueous salt solutions (NaCl)	S
Polar solvents (C_2H_5OH)	S
Nonpolar solvents (C_6H_6)	Q
Water	S
Aqueous alkaline solutions (NaOH)	S

*lb/(in^2 · 0.145) = KPa (kilopascals); ft · lb/(in of notch · 0.0187) = cm · N/cm of notch.

†S = satisfactory; Q = questionable; U = unsatisfactory.

Because of their excellent electrical properties, radiation resistance, and high dimensional stability, up to 204°C (400°F), DAP polymers are used for the compression molding of electrical connectors and insulators. Fiberglass-reinforced DAP is used in aircraft and missile components, as well as in radomes and tubular ducts. Specific thin, allyl-molded sections have a UL rating of 94V-O. They also have a high arc ignition and track resistance.

Scratch-resistant lenses are produced by the vaporization of silica or alumina coatings on the surface of the polymer, which is produced by high-energy-radiation polymerization of DAP in lens-shaped molds. The name "allyl" is derived from the Latin word *allium,* meaning "garlic." In spite of the obnoxious odor of diallyl esters, the moldings are odorless, colorless, and nontoxic.

ASTM 1636 specification classifies DAP composites as type I, glass-fiber reinforced; type II, mineral-filled; and type III, synthetic-fiber reinforced. There are also subclassifications for the ortho and iso (meta) phthalates. The two carboxyl groups are on adjacent carbon atoms (1 and 2) in the ortho isomer and on alternate carbon atoms (1 and 3) in the meta isomer.

3.7 Diallyl Diglycol Carbonate

Diallyl diglycol carbonate (DADC), which is produced by the condensation of phosgene ($COCl_2$) and allyl alcohol ($CH_2{=}CHCH_2OH$) in the presence of sodium hydroxide, may be polymerized in a manner similar to that described for DAP. This clear thermoset polymer (CR 39) is usually produced by creating a prepolymer syrup in a glass cell.

The original use of CR-39 was as windows for aircraft during World War II, but the principal applications today are as impact-resistant lenses, safety shields, and solid-state track detectors (SSTDs) for the evaluation of nuclear particles. Tracks, which may be made visible by partial saponification (hydrolytic degradation) of CR-39, are formed when CR-39 sheets are subjected to radiation. The scratch resistance of these lenses is enhanced by vapor deposition, or sputtering, of silicon oxide by electron beams or glow discharge.

The monomer used for CR-39 is a nontoxic colorless liquid which may cause dermatitis if it comes in contact with the skin. Allyl alcohol produced by the hot alkaline saponification of DADC is toxic. The limiting dose of allyl alcohol in rats is 64 milligrams per kilogram (mg/kg), i.e., 64 parts in 1,000,000 parts body weight.

The properties of a typical cast polymer of CR-39 are given in Table 3.3.

3.8 Alkyd Resins

Many investigators modified Watson Smith's original Glyptal resins by replacing moderate amounts of the difunctional phthalic anhydride by monobasic acids, such as oleic acid ($C_{17}H_{33}COOH$), and R. Kienle perfected these modifications by the production of alkyds. Alkyds, which are the most widely used coatings, are also used as molded thermosets.

Leo Baekeland employed the concept of functionality f in the production of novolac PF resins, and W. H. Carothers refined these concepts and produced linear polymers by the condensation of difunctional acids [$R(COOH)_2$] with difunctional alcohols [$R(OH)_2$] and difunctional amines [$R(NH_2)_2$].

Carothers showed that unusually high purity of the reactants was

TABLE 3.3 Properties of a Typical Polymer of DADC

Property, units*	
Melting point T_m, °F	—
Glass transition temp. T_g, °F	—
Processing temp., °F	—
Molding pressure, 10^3 lb/in^2	—
Mold shrinkage, 10^{-3} in/in	—
Heat deflection temp. under flexural load of 264 lb/in^2, °F	160
Maximum resistance to continuous heat, °F	150
Coefficient of linear expansion, 10^{-6} in/(in · °F)	100
Compressive strength, 10^3 lb/in^2	21
Izod impact strength, ft · lb/in of notch	—
Tensile strength, 10^3 lb/in^2	5
Flexural strength, 10^3 lb/in^2	9
Percent elongation	1
Tensile modulus, 10^3 lb/in^2	300
Flexural modulus, 10^3 lb/in^2	300
Rockwell hardness	M95
Specific gravity	1.3
Percent water absorption	0.2
Dielectric constant	—
Dielectric strength, V/mil	380
Resistance to chemicals at 75°F:†	
Nonoxidizing acids (20% H_2SO_4)	S
Oxidizing acids (10% HNO_3)	Q
Aqueous salt solutions (NaCl)	S
Polar solvents (C_2H_5OH)	S
Nonpolar solvents (C_6H_6)	Q
Water	S
Aqueous alkaline solutions (NaOH)	S

*lb/(in^2 · 0.145) = KPa (kilopascals); ft · lb/(in of notch · 0.0187) = cm · N/cm of notch.

†S = satisfactory; Q = questionable; U = unsatisfactory.

essential for the production of linear high-molecular-weight polymers. According to the Carothers' equation, Eq. (3.8), the extent of reaction, or fractional yield, p, for condensation polymerization must approach 1.0 for the production of high-molecular-weight polymers as measured by the degree of polymerization DP, which is equal to the number of repeating units in the polymer molecule. Since the product consists of many molecules of different molecular weights, the DP is shown as an average DP, or $\overline{\mathrm{DP}}$.

$$\overline{\mathrm{DP}} = \frac{1}{1-p} \tag{3.8}$$

The strength of a polymer is dependent, to a large extent, on chain entanglement, which does not occur until the DP is about 100. Lower

DP values are acceptable when strong intermolecular forces, such as dipole-dipole interaction and hydrogen bonding, are present. (The attraction between a chlorine atom on one polymer chain and a hydrogen atom on another chain is called a dipole-dipole interaction. The attraction between an oxygen or nitrogen atom on one polymer chain and a hydrogen atom on another chain is called hydrogen bonding.)

Carothers also showed that when one of the reactants is trifunctional, the DP becomes infinite at the gel point. Gelation cannot occur and linear polymers are produced, regardless of the magnitude of the P value, when the f of the reactants is not greater than 2.

As was discussed qualitatively, many investigators, prior to the research in the early 1930s by Carothers and Kienle, recognized that f can be reduced by the incorporation of monofunctional compounds in the reactant mix.

Alkyds are now produced by the condensation of glycerol, fatty acids, and phthalic anhydride. The alkyd resins are classified as short, medium, long, and very long oil lengths with a percentage fatty acid content of less than 42 percent, less than 54 percent, less than 68 percent, and greater than 68 percent, respectively.

Alkyds may be thermoplastic if f is 2, as in the case of difunctional ethylene glycol and difunctional phthalic anhydride, However, most alkyds, including those used for protective coatings, are made from glycerol, which has an f of 3, and hence are thermoset. Pentaerythritol [$(HOCH_2)_4C$], which has an f of 4, may also be used. It is customary to heat phthalic anhydride and glycerol at 179°C (355°F) to produce a syrup (first stage) and to add molten fatty acids to form esters (RCOOR′) by reaction with the residual hydroxyl groups in the syrup.

Drying (curable) alkyds are characterized by specific drying times, which are related to the amount of fatty acid added to the reactants in the production of the alkyd. It is customary to add lead or cobalt salts of naphthenic or octoic acid as driers (initiators) in order to produce good paint films. ("Naphthenic acid" is a term used to describe nonparaffinic acids obtained during the distillation of crude oil.)

The properties of a typical molded alkyd are shown in Table 3.4.

Over 130,000 t of molded alkyd plastics was consumed in the United States in 1987. Molded alkyds are used in applications in which dimensional stability, good electrical properties, and better-than-average mechanical properties are required. The colored molded alkyd plastics are used for appliances. Alkyds are transparent to microwave applications and are also used as circuit breakers, transformer housings, distributor caps, and rotors. Specialty alkyds are able to withstand the temperature of vapor-phase soldering without loss of integrity.

While some individuals are sensitive to some of the reactants and

TABLE 3.4 Properties of a Typical Molded Alkyd

Property, units*	Mineral-filled	Fiberglass-reinforced
Melting point T_m, °F	—	—
Glass transition temp. T_g, °F	—	—
Processing temp., °F	300	300
Molding pressure, 10^3 lb/in^2	15	20
Mold shrinkage, 10^{-3} in/in	6	5
Heat deflection temp. under flexural load of 264 lb/in^2, °F	400	425
Maximum resistance to continuous heat, °F	350	375
Coefficient of linear expansion, 10^{-6} in/(in · °F)	35	25
Compressive strength, 10^3 lb/in^2	25	27
Izod impact strength, ft · lb/in of notch	0.4	2
Tensile strength, 10^3 lb/in^2	5	15
Flexural strength, 10^3 lb/in^2	1.2	1
Percent elongation	—	—
Tensile modulus, 10^3 lb/in^2	2000	2500
Flexural modulus, 10^3 lb/in^2	2000	2000
Rockwell hardness	E98	E95
Specific gravity	2.0	2.0
Percent water absorption	0.2	0.2
Dielectric constant	—	—
Dielectric strength, V/mil	400	450
Resistance to chemicals at 75°F:†		
Nonoxidizing acids (20% H_2SO_4)	S	S
Oxidizing acids (10% HNO_3)	U	U
Aqueous salt solutions (NaCl)	S	S
Polar solvents (C_2H_5OH)	S	S
Nonpolar solvents (C_6H_6)	U	U
Water	S	S
Aqueous alkaline solutions (NaOH)	Q	Q

*lb/(in^2 · 0.145) = KPa (kilopascals); ft · lb/(in of notch · 0.0187) = cm · N/cm of notch.
†S = satisfactory; Q = questionable; U = unsatisfactory.

solvents used in the production of alkyds, they are not affected by the cured alkyd resins, which are nontoxic. Alkyd coatings are characterized by rapid drying, good adhesion, flexibility, durability, and mar resistance. Alkyds may be modified by blending with cellulose nitrate, amino resins, chlorinated rubber, phenolic resins, polyamides, epoxy resins, and silicones. Since the alkyds are unsaturated, they may be copolymerized with styrene, vinyltoluene, *p*-methylstyrene, methyl methacrylate, or acrylonitrile.

Reinforced unsaturated polyesters are discussed in Chap. 4.

3.9 Glossary

acrylates Polymers with ester pendant groups, such as polymethyl methacrylate, $\{CH_2C(CH_3)COOCH_3\}$.

addition polymerization A polymer-producing reaction in which an active species, such as a free radical (R ·), adds to a vinyl monomer and the product adds sequentially to other monomer molecules.

A-glass High-alkali glass.

alkyds Polyesters with unsaturated groups.

allyl group $CH_2{=}CH{-}CH_2{-}$.

aramid An aromatic polyamide.

ASTM American Society for Testing and Materials.

boron B.

boron filaments Composite filaments produced by the deposition resulting from the chemical vapor deposition of the reaction of boron trichloride (BCl_3) and hydrogen (H_2) on a tungsten or graphite core.

carbonization Pyrolysis.

carboxyl group COOH.

Carothers, W. H. A pioneer polymer chemist who developed the equation for predicting DP and gelation in condensation polymerization.

Carothers' equation $\overline{DP} = 1/(1-p)$, where p = extent of reaction.

catalyst A substance which changes the rate of reaction without undergoing permanent change in composition.

Celluloid Plasticized cellulose nitrate.

cellulose nitrate A compound produced by the reaction of cellulose and nitric acid.

chain reaction One which includes initiation, propagation, and termination steps.

chain-reaction polymerization Addition polymerization.

chain-transfer agent A compound with a labile atom or group which can be readily abstracted by a macroradical.

coefficient of expansion A fractional change in dimensions of a test bar, per degree of temperature.

compressive strength The ability of material to resist a crushing force.

condensation polymerization A polymer-producing reaction in which a functional group, in a difunctional molecule, condenses with a functional group in another difunctional molecule and this condensation continues, stepwise, to produce a polymer.

coupling Termination by the combination of two macroradicals.

CR-39 A commercial diallyl polymer (DADC).

cross-linked polymer A three-dimensional (network) thermoset polymer.

D Diameter.

DADC Diethylene glycol bischloroformate or diallyl diglycol carbonate.

DAP Diallyl phthalate.

dead polymer A polymer obtained after the termination of propagation.

dermatitis Irritation and possible blistering of skin.

diallyl phthalate $C_6H_4(COOCH_2CH{=}CH_2)_2$ (benzene ring with two adjacent $COOCH_2CH{=}CH_2$ groups)

dielectric constant A measure of electric charge stored per unit volume between plates at unit potential.

difunctional Having two functional groups, such as ethylene glycol ($HOCH_2CH_2OH$).

DP Degree of polymerization, the number of repeating units in a polymer molecule.

drier A heavy metal salt, such as cobalt naphthenate.

drying The cross-linking of oleoresinous paint or alkyd in the presence of air and a drier (catalyst).

E Modulus.

E_{fu} The ultimate breaking strength of a fiber.

E-glass A borosilicate glass.

ejection The removal of a molded piece from the mold.

elongation at break The percent elongation at the moment of rupture of a test specimen.

ethylene glycol $HOCH_2CH_2OH$.

f Functionality.

fiber A relatively short length of a filament, with a minimum length/diameter ratio of 100.

fiberglass Filaments of glass obtained by forcing molten glass through small holes (bushings).

filament A continuous fiber.

flexural strength The resistance of a test specimen to bending stress.

free radical An electron-deficient molecule.

functionality The number of reactive (functional) groups in a molecule.

gel point The stage in a polymerization at which cross-linking begins.

glycerol $HOCH_2CHOHCH_2OH$.

Glyptal A pioneer thermoset polyester produced by the condensation of glycerol and phthalic anhydride.

graphitization The production of graphite by pyrolysis of carbon at elevated temperatures (>1800°C) in an inert atmosphere.

hardness Resistance to surface indentation.

heat deflection temperature DTUL, formerly called heat distortion temperature; the temperature at which a standard bar deflects a specific distance under a stated flexural load.

hydrolysis Decomposition in the presence of water.

impact strength The ability of a test specimen to resist fracture when subjected to a hammerlike blow.

initiator An active species, such as a free radical, which initiates chain-reaction polymerization.

isophthalic acid meta-Phthalic acid.

Izod impact test The standard ASTM destructure impact test, in which the specimen usually includes a notch, with specified indentations.

K Alkyd constant derived by Patton.

Kelley-Tyson One of the prevalent concepts of estimating stress buildup of reinforced fibers.

Kevlar An aramid: poly-*p*-terephthalamide.

Kienle, R. H. The inventor and coiner of the name "alkyd resin."

L Load.

l_c Critical fiber lengths.

laminate A composite consisting of layers of resin and reinforcing sheets.

law of mixtures Properties of a mixture rule that are related to the relative amounts of additives present in the mixture.

macroradical A high-molecular-weight radical.

meta In 1,3 positions on the benzene ring, i.e.,

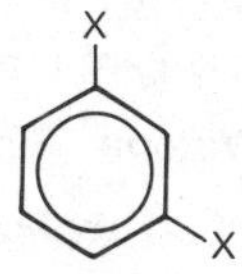

modulus The ratio of stress to strain.

mold The cavity in which the prepolymer is placed, where it assumes the shape of the cavity, usually at a high temperature and pressure.

monomer The building unit in addition polymerization.

oil length A qualitative measure of the relative amounts of unsaturated oil used in the production of an alkyd resin.

oleic acid An unsaturated high-molecular-weight acid ($C_{17}H_{33}COOH$).

oligomer A very-low-molecular-weight polymer.

organosilane An R_4Si coupling agent.

organotitanate A coupling agent based on titanium compounds.

organozirconate A coupling agent based on zirconium compounds.

ortho In adjacent positions on the benzene ring.

p The extent of reaction.

peroxide initiator An organic peroxide, such as benzoyl peroxide: $(C_6H_5CO_2)_2$.

phase, continuous A resin matrix in a composite.

phase, discontinuous A reinforcing member of a composite.

phthalate An ester of phthalic acid.

phthalic anhydride $C_6H_4C_2O_3$.

phosgene $COCl_2$.

polyester A polymer with ester groups $\leftarrow COOR \rightarrow$, usually in the polymer backbone.

promoter An accelerator or catalyst.

propagation The propagation chain growth addition polymerization.

R · A free radical.

radical chain polymerization One in which the initiating species is a free radical (R ·).

Rockwell hardness A number related to the depth of penetration of an indenter, under a specified load.

roving A collection of bundles of filaments.

S-glass Magnesia-alumina silicate glass which is used to produce filaments with high tensile strength.

shear Applied force which tends to cause one surface to pass by another as in the buttering of a piece of toast.

Smith, Watson The inventor of Glyptal resins.

spinneret Tiny holes through which a resin is forced in the production of filaments.

staple fiber A short-length filament.

strain The rate of change in length to the original length, i.e., percent elongation.

strand A bundle of continuous filaments gathered together as a single unit.

strength, flexural The resistance of a test specimen to bending under load.

stress The rate of applied load to cross-sectional area.

tensile strength The resistance to a pulling force.

thermoplastic A polymer which can be reversibly softened by heat and hardened by cooling.

thermoset An infusible, insoluble polymer which cannot be softened by heat.

trifunctional Having three functional groups, such as glycerol ($H_2COHCHOHCH_2OH$).

unsaturated polyester A polyester prepolymer with ethylenic (C=C) groups in the backbone.

V Volume.

vinyl compound $H_2C{=}HX$.

viscosity Resistance to flow.

Volan A methacrylatochromic chloride coupling agent.

wrap angle The angle at which filament is wound on the mandrel in the filament-winding process.

3.10 References

Harper, C. E. (ed.): *Handbook of Plastics and Elastomers,* Van Nostrand Reinhold Company, Inc., New York, 1975.

Lanson, H. J.: in J. I. Kroschwitz (ed.), *Encyclopedia of Polymer Science and Engineering*, vol. 2, John Wiley & Sons, Inc., Interscience Publishers, New York, 1985.

Moore, G. R., and E. E. Kline: *Properties of Processing of Polymers for Engineers,* Prentice-Hall, Inc., Englewood Cliffs, N.J., 1984.

Morgans, W. M.: *Outlines of Paint Technology,* Charles Griffin & Company, Ltd., High Wycombe, U.K., 1982.

Patton, T. C.: *Alkyd Resin Technology,* John Wiley & Sons, Inc., Interscience Publishers, New York, 1962.

Schildknecht, C. E.: *Allyl Compounds and Their Polymers,* John Wiley & Sons, Inc., Interscience Publishers, New York, 1973.

Schildknecht, C. E.: in J. I. Kroschwitz (ed.), *Encyclopedia of Polymer Science and Engineering*, vol. 4, John Wiley & Sons, Inc., Interscience Publishers, New York, 1986.

Seymour, R. B.: *Plastics for Engineering Applications,* AMS International, Metals Park, Ohio, 1986.

Seymour, R. B.: *Introduction to Polymer Chemistry,* McGraw-Hill Book Company, New York, 1971.

Seymour, R. B.: *Modern Plastics Technology,* Reston Publishing Co., Inc., Reston, Va., 1975.

Seymour, R. B.: *Plastics vs. Corrosives,* John Wiley & Sons, Inc., Interscience Publishers, New York, 1982.

Tomalia, D. A., and G. R. Killat: in J. I. Kroschwitz (ed.), *Encyclopedia of Polymer Science and Engineering*, vol. 2, John Wiley & Sons, Inc., Interscience Publishers, New York, 1985.

Chapter 4

Unsaturated Polyester, Epoxy, and Polyimide Resins

4.1 Introduction

Unsaturated polyesters, such as polyethylene maleate, which can be converted to cross-linked polymers by heating with benzoyl peroxide, were invented by Carleton Ellis in the mid-1930s. He simplified the fabrication of these prepolymers by dissolving them in vinyl acetate before the addition of the peroxide initiator. Because of the high volatility of vinyl acetate, he replaced this monomer with styrene, and this system continues to be used today.

4.2 Unsaturated Polyester (Polyethylene Maleate) Resin

The unsaturated polyester prepolymer is produced by the step-reaction polymerization (condensation) of ethylene glycol ($HOCH_2CH_2OH$) and maleic anhydride. It is now known that the cis configuration of this reactant is converted to a trans configuration during the condensation and that the trans isomer polymerizes (cures faster) than the cis isomer.

In the cis configuration, the functional groups are on the same side of the ethylenic double bond, i.e.,

$$\overset{|}{C}{=}\overset{|}{C}$$

and these groups are on alternate sides in the trans configuration,

$$\overset{|}{C}{=}\underset{|}{C}$$

Since maleic anhydride is a cyclic molecule, it can exist only in the cis configuration. However, the acid formed by the hydrolysis of the anhydride can exist in either form. The trans form is called fumaric acid.

The rate of addition polymerization may be accelerated and the occurrence of a sticky surface may be prevented by the addition of a small amount of a drier, such as cobalt naphthenate. Acceleration of the rate of polymerization also occurs when traces of tertiary amines are added. Wax is sometimes added to the prepolymer system to prevent the formation of a sticky surface in the presence of air. The accelerators are sometimes called promotors.

Unsaturated polyesters are used, to a small extent (1%), as coatings, but their principal use is as matrices for reinforced plastics (RPs). The principal type is fiberglass-reinforced plastic (FRP). Fiberglass was fortuitously commercialized in the late 1930s, coincidental with the development of brittle unsaturated polyester resins.

Fiber reinforcements, such as graphite, aramid, and blends (hybrids) of fibers are also used for reinforcing polyesters and other plastics. The principal fabrication techniques are "hand lay-up" or "spray lay-up," but these labor-intensive techniques are being replaced by more sophisticated fabrication. The fabrication methods described for glass-reinforced polyesters (GRPs) are also applicable to other polymeric systems.

The condensation reaction product of maleic anhydride and ethylene glycol is shown in Eq. (4.1):

$$HO(CH_2)_2OOCCH{=}CHCOOH \tag{4.1}$$

Since polyethylene maleate is produced by step-reaction polymerization, its average degree of polymerization ($\overline{DP}$) is governed by Carothers' equation, Eq. (3.8), i.e., $\overline{DP} = 1/(1 - p)$, where p is the extent of reaction, or fractional yield.

In contrast, the copolymerization of styrene and the unsaturated polyester is a radical chain reaction which follows the initiation, propagation, and termination steps described for chain polymerization in Chap. 3. It is customary to use about 30 percent styrene and 70 percent of the polyester prepolymer and to initiate the copolymerization by the addition of an organic peroxide, such as benzoyl peroxide.

It is advantageous to use the initiator as a "phlegmatized" paste, i.e., a mixture of plasticizer and initiator. It is also customary to add a finely divided, partially soluble thermoplastic, such as PS or PMMA (low-profile additives) to reduce shrinkage during polymerization and thus provide a smoother surface.

The properties of a typical glass-reinforced unsaturated polyester composite are shown in Table 4.1.

TABLE 4.1 Properties of Typical Fiberglass-Reinforced Polyesters

Property, units*	Sheet-molding compound (SMC)	Bulk-molding compound (BMC)
Melting point T_m, °F	—	—
Glass transition temp. T_g, °F	—	—
Processing temp., °F	300	300
Molding pressure, 10^3 lb/in^2	1	1
Mold shrinkage, 10^{-3} in/in	3	4
Heat deflection temp. under flexural load of 264 lb/in^2, °F	425	375
Maximum resistance to continuous heat, °F	375	325
Coefficient of linear expansion, 10^{-6} in/(in · °F)	15	15
Compressive strength, 10^3 lb/in^2	20	20
Izod impact strength, ft · lb/in of notch	15	7.5
Tensile strength, 10^3 lb/in^2	15	10
Flexural strength, 10^3 lb/in^2	25	20
Percent elongation	3	2
Tensile modulus, 10^3 lb/in^2	2000	2000
Flexural modulus, 10^3 lb/in^2	1500	1500
Rockwell hardness	M50	M50
Specific gravity	2.0	2.0
Percent water absorption	0.2	0.3
Dielectric constant	5	5
Dielectric strength, V/mil	450	350
Resistance to chemicals at 75°F:†		
Nonoxidizing acids (20% H_2SO_4)	Q	Q
Oxidizing acids (10% HNO_3)	U	U
Aqueous salt solutions (NaCl)	S	S
Polar solvents (C_2H_5OH)	Q	Q
Nonpolar solvents (C_6H_6)	U	U
Water	S	S
Aqueous alkaline solutions (NaOH)	U	U

*lb/(in^2 · 0.145) = KPa (kilopascals); ft · lb/(in of notch · 0.0187) = cm · N/cm of notch.
†S = satisfactory; Q = questionable; U = unsatisfactory.

4.3 Other Unsaturated Polyester Resins

In addition to phthalic anhydride, highly chlorinated derivatives of phthalic anhydride, such as tetrachlorophthalic anhydride and the related chlorendic anhydride, are used to produce flame-resistant composites. Chlorendic anhydride ($C_9H_2Cl_6O_3$) is also called hexachloroendomethylenetetrahydrophthalic anhydride (HET). More alkaline-resistant unsaturated polyesters are also produced from isophthalic acid, from bisphenol A and fumaric acid, and from vinyl ester, which is the ester of bisphenol A, propylene oxide, and methacrylic acid.

Because of low cost, most general-purpose unsaturated polyester prepolymers are produced by the esterification of maleic anhydride in the presence of phthalic anhydride or *o*-phthalic acid. However, better mechanical properties of the cured composites are produced when isophthalic (meta) acid is used. It is customary to dissolve these prepolymers in styrene. However, the concentration of volatile organic compounds (VOCs) is reduced when vinyltoluene or *p*-methylstyrene is used in place of styrene. The term "vinyltoluene" is used to describe a mixture of meta and para methyl-substituted styrenes.

4.4 Initiators (Catalysts) for Curing Unsaturated Polyesters

The most widely used initiators for the polymerization of styrene in the prepolymer solution are methyl ethyl ketone peroxide (MEKP) and benzoyl peroxide (BPO). It is customary to add an activator, such as *N,N*-dimethylaniline, when the prepolymer system is cured at room temperature. It is also customary to include a small amount of cobalt naphthenate with the tertiary amine accelerator.

4.5 Vinyl Ester

An unsaturated polyester, produced by the polymerization of the ester formed from the adduct of propylene oxide and bisphenol A with methacrylic acid (Derakane, Epocryl, Corezin, Corrolite), is also available. These resins are usually dissolved in styrene and polymerized like conventional unsaturated polyesters. The vinyl ester resins are more resistant to corrosion and shrink less than general-purpose polyesters.

The properties of a typical vinyl ester resin are shown in Table 4.2.

4.6 Epoxy Resins

Most epoxy (EP) resins are prepared by the reaction of epichlorohydrin (CH_2OCHCH_2Cl) and bisphenol A [BPA; *p,p'*-isopropylidenephenol; $HOC_6H_4C(CH_3)_2C_6H_4OH$]. BPA is the product of the condensation of acetone and phenol. However, some epoxy resin precursors are produced by the reaction of peracetic acid (H_3CCO_3H) and cyclohexene.

Epoxy prepolymers contain terminal epoxy

(—C(O)C—)

groups which react with amines at room temperature and many pendant hydroxyl groups which react with cyclic anhydrides at elevated temperatures. Over 183,000 t of EP was used in the United States in

TABLE 4.2 Properties of a Typical Vinyl Ester Resin

Property, units*	Unfilled	25% Fiberglass
Melting point T_m, °F	—	—
Glass transition temp. T_g, °F	—	—
Processing temp., °F	—	—
Molding pressure, 10^3 lb/in^2	—	—
Mold shrinkage, 10^{-3} in/in	—	—
Heat deflection temp. under flexural load of 264 lb/in^2, °F	200	200
Maximum resistance to continuous heat, °F	175	175
Coefficient of linear expansion, 10^{-6} in/(in · °F)	20	15
Compressive strength, 10^3 lb/in^2	17	30
Izod impact strength, ft · lb/in of notch	12	10
Tensile strength, 10^3 lb/in^2	15	20
Flexural strength, 10^3 lb/in^2	20	20
Percent elongation	6	3
Tensile modulus, 10^3 lb/in^2	2000	2000
Flexural modulus, 10^3 lb/in^2	1200	1200
Rockwell hardness	M85	M105
Specific gravity	—	—
Percent water absorption	—	—
Dielectric constant	0.2	0.2
Dielectric strength, V/mil	5	5
Resistance to chemicals at 75°F:†		
Nonoxidizing acids (20% H_2SO_4)	S	S
Oxidizing acids (10% HNO_3)	Q	Q
Aqueous salt solutions (NaCl)	S	S
Polar solvents (C_2H_5OH)	S	S
Nonpolar solvents (C_6H_6)	Q	Q
Water	S	S
Aqueous alkaline solutions (NaOH)	Q	Q

*lb/(in^2 · 0.145) = KPa (kilopascals); ft · lb/(in of notch · 0.0187) = cm · N/cm of notch.
†S = satisfactory; Q = questionable; U = unsatisfactory.

1987, but over 50 percent of this volume was used as embedding media, adhesives, and coatings. About 20 percent of this volume was used to produce high-grade reinforced polymeric composites which can be molded and fabricated by techniques described for unsaturated polyesters.

Because of the presence of pendant hydroxyl groups, epoxy resins bond well to other materials. Molded or cast epoxy resins, which have excellent dimensional stability and low shrinkage, are used as dies for stamping metal sheet and as models for production articles.

Many different polyamines, such as diethylenetriamine [DETA; $H_2N(CH_2)_2NH(CH_2)_2NH_2$] will cure liquid epoxy prepolymers at room temperature. Since these reactants are toxic and skin irritants, adequate ventilation must be provided and the skin of the applicator must be protected from contact with the curing agents to prevent dermatitis. Phthalic anhydride is widely used as a curing agent at elevated temperatures. Flame-retardant resins are produced when HET is used as a curing agent. EP syntactic foams are produced when hollow glass spheres are added to the resin mix. Properties of typical epoxy resins and composites are shown in Table 4.3. The formula for EP is shown in Fig. 4.1.

4.7 Thermosetting PIs

PIs are produced by the condensation of an aromatic dianhydride, such as mellitic dianhydride $[(C_2O_3)_2C_6H_2(C_2O_3)_2]$, and an aromatic

TABLE 4.3 Properties of Typical Epoxy Resins

Property, units*	Unfilled EP	Glass-sphere-filled EP	Fiberglass-reinforced EP
Melting point T_m, °F	—	—	—
Glass transition temp. T_g, °F	—	—	—
Processing temp., °F	—	275	300
Molding pressure, 10^3 lb/in^2	—	0.5	2.5
Mold shrinkage, 10^{-3} in/in	—	8	5
Heat deflection temp. under flexural load of 264 lb/in^2, °F	225	225	350
Maximum resistance to continuous heat, °F	200	200	300
Coefficient of linear expansion, 10^{-6} in/(in · °F)	25	—	20
Compressive strength, 10^3 lb/in^2	20	12	30
Izod impact strength, ft · lb/in of notch	0.5	0.2	2
Tensile strength, 10^3 lb/in^2	10	3	12
Flexural strength, 10^3 lb/in^2	16	6	20
Percent elongation	4	—	4
Tensile modulus, 10^3 lb/in^2	350	—	3000
Flexural modulus, 10^3 lb/in^2	—	600	3000
Rockwell hardness	M90	—	M105
Specific gravity	1.25	0.85	1.8
Percent water absorption	0.1	0.4	0.2
Dielectric constant	4	4	4
Dielectric strength, V/mil	400	400	300
Resistance to chemicals at 75°F:†			
Nonoxidizing acids (20% H_2SO_4)	S	S	S
Oxidizing acids (10% HNO_3)	U	U	U
Aqueous salt solutions (NaCl)	S	S	S
Polar solvents (C_2H_5OH)	S	S	S
Nonpolar solvents (C_6H_6)	S	S	S
Water	S	S	S
Aqueous alkaline solutions (NaOH)	S	S	S

*lb/(in^2 · 0.145) = KPa (kilopascals); ft · lb/(in of notch · 0.0187) = cm · N/cm of notch.
†S = satisfactory; Q = questionable; U = unsatisfactory.

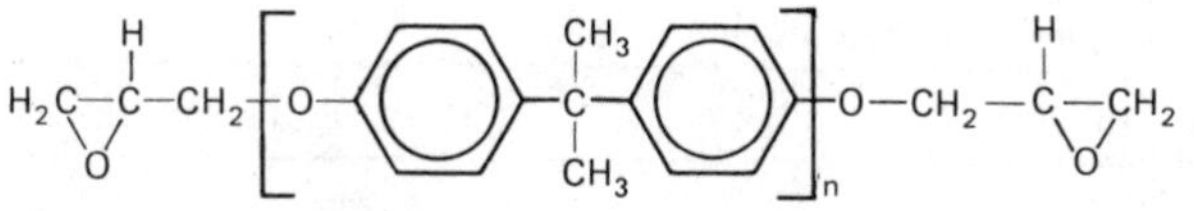

Figure 4.1 EP polymer.

diamine, such as methylenedianiline]$(H_2NC_6H_4)_2C(CH_3)_2$]. The intermediate polyamic acid, which is produced in the first step of the condensation, is imidized (dehydrocyclized) in the presence of a dehydrating agent, such as acetic anhydride. The precursor polyamic acid is marketed by Monsanto as Skybond Resin and by Du Pont as Pyralin.

Unfilled PI is used as a heat-resistant film (Kapton) and wire enamel. Fiberglass-reinforced PI is used for jet engines and aircraft turbine engine parts. Rhone-Poulenc markets its PI laminates under the trade name Keramid, and Du Pont markets its PI moldings under the trade name Vespel. The properties of a fiberglass-filled PI composite are shown in Table 4.4.

Bismaleimides (BMIs), which are more readily processed than PI, are produced by the condensation of maleic anhydride and a diamine. The resulting prepolymers are cured at elevated temperatures and used as high-temperature-resistant coatings and aircraft parts.

4.8 Properties of Composites

The physical properties of a unidirectional composite are comparable to those of wood, i.e., the modulus has a high value along the grain, but this value is much less when measured perpendicular to the grain. In either case, it is more realistic to report these properties on a volume rather than on a weight basis. These specific property values are obtained by dividing the weight-biased values by the density. The density of steel is about 4 times that of a polymeric composite.

The properties of composites containing randomly oriented reinforcements are controlled by two independent constants, namely, Young's modulus and the Poisson ratio. Young's modulus is the ratio of applied tension stress to the resulting strain, parallel to the tension. The Poisson ratio is the ratio of the transverse contracting strain to the elongation strain as measured on a rod stretched by forces at the ends of the rod, which are parallel to the rod's axis.

The properties of the more ordered (isotropic) composites require at least two Young's moduli and two Poisson ratios. For reasons of simplicity, one considers the internal stresses and strains as those in a heterogeneous system. This idealized approach, based on the law of mixtures, provides a semiquantitative framework for predicting the behavior of composites.

TABLE 4.4 Properties of a Typical Polyimide Composite

Property, units*	50% Fiberglass-filled PI
Melting point T_m, °F	—
Glass transition temp. T_g, °F	—
Processing temp., °F	425
Molding pressure, 10^3 lb/in^2	6
Mold shrinkage, 10^{-3} in/in	2
Heat deflection temp. under flexural load of 264 lb/in^2, °F	660
Maximum resistance to continuous heat, °F	600
Coefficient of linear expansion, 10^{-6} in/(in · °F)	7
Compressive strength, 10^3 lb/in^2	34
Izod impact strength, ft · lb/in of notch	5
Tensile strength, 10^3 lb/in^2	6.5
Flexural strength, 10^3 lb/in^2	21
Percent elongation	1
Tensile modulus, 10^3 lb/in^2	—
Flexural modulus, 10^3 lb/in^2	2000
Rockwell hardness	M118
Specific gravity	1.7
Percent water absorption	0.7
Dielectric constant	—
Dielectric strength, V/mil	450
Resistance to chemicals at 75°F:†	
Nonoxidizing acids (20% H_2SO_4)	Q
Oxidizing acids (10% HNO_3)	Q
Aqueous salt solutions (NaCl)	S
Polar solvents (C_2H_5OH)	S
Nonpolar solvents (C_6H_6)	S
Water	S
Aqueous alkaline solutions (NaOH)	U

*lb/(in^2 · 0.145) = KPa (kilopascals); ft · lb/(in of notch · 0.0187) = cm · N/cm of notch.
†S = satisfactory; Q = questionable; U = unsatisfactory.

Thus, one may show that the properties of a composite, P, are equal to the sum of the properties of the components multiplied by their fractional volumes V as shown in Eq. (4.2).

$$P = P_1V_1 + P_2V_2 + P_3V_3 + \cdots + P_nV_n \tag{4.2}$$

Thus, the heat capacity and density of a dense composite can be calculated from this rule of mixtures. The Young's modulus E_{11} of a continuous-fiber-reinforced composite, in the direction of the fiber, is also equal to or greater than the modulus of the fiber, E_f, and the modulus of the resin matrix, E_m, multiplied by their fractional volumes as shown in Eq. (4.3).

$$E_{11} \geq V_fE_f + V_mE_m \tag{4.3}$$

If the reinforcing fibers are initially discontinuous or are fractured, V_f must be multiplied by a constant ζ, which is dependent on the geometry and elastic deformation properties of the fiber, the matrix, and their interface. Accordingly, E_{22}, i.e., Young's modulus transverse to the fiber axis, is dependent on the viscosity of the matrix, η, and a constant ζ based on the geometry of the fiber as shown in Eqs. (4.4) and (4.5).

$$E_{22} \approx E_m \frac{1 + \zeta \eta V_f}{1 - \eta V_f} \tag{4.4}$$

where

$$\eta = \frac{E_f - E_m}{E_f + \zeta E_m} \tag{4.5}$$

The major Poisson ratio ν_{12} and the minor Poisson ratio ν_{21} are shown in Eqs. (4.6) and (4.7).

$$\nu_{12} \approx V_f \nu_f + V_m \nu_m \tag{4.6}$$

$$\nu_{21} = \frac{\nu_{12} E_{12}}{E_{11}} \tag{4.7}$$

The strength of a composite, S, may also be approximated from the rule of mixtures as shown by Eq. (4.8), in which σ^* is the stress in the matrix at failure strain and $\overline{S}_f$ is the mean fiber strength.

$$S_{\text{composite}} = V_f \overline{S}_f + V_m \sigma^* \tag{4.8}$$

The value of σ^* is somewhat subjective and dependent upon one's criteria for the failure mode. Composites may fail by cracking, and this may be prevented by good adhesive bonding at the interface. Failure may also be related to the presence of voids in the composite. Buckling may be initiated by misalignment or heterogeneous distribution of the resin throughout the composite. Sound fabrication techniques are also important in preventing failure.

4.9 Bulk Molding

Composites (BMCs) may be produced by the bulk-molding process, in which a slurry of resin, initiator (catalyst), mold release agents, chopped fibers, filler, and pigment are mixed in a high-shear mixer and the resultant molding compound is extruded as a rope or log. This dough, which is sometimes called a dough-molding compound, is then compression-, transfer-, or injection-molded, using techniques described in Chap. 2. It is customary to use a telescoping compression mold in which the edges between the male and female die are sealed before the application of pressure to the mold. Matched metal dies are used when close tolerances of the final part are specified.

4.10 Sheet Molding

Composites (SMCs) may also be produced by the sheet-molding process, in which a pastelike mixture of catalyzed prepolymer is spread, by use of a doctor's blade, onto two polyethylene sheets to which chopped fibers are added. The composite sheet is compacted by cold rollers, followed by heated rollers. The sheets are allowed to age (thicken) and are then cut to the required size and shape, and the reinforced sheet is compression-molded. In addition to the standard SMC process, a thick-molding (TMC) and a reinforced X-pattern sheet (XMC) may be used as improved modifications of SMC.

4.11 Hand Lay-Up

The simplest and crudest fabrication technique for making reinforced composite structures is the crude hand lay-up procedure. The tooling for this fabrication may be wood, plastic, or metal in the form of the desired part. The surface of the crude mold is coated with wax, which serves as a parting agent, before the application of a nonreinforced slow-curing prepolymer (gelcoat).

After the preform has set (partially polymerized), layers of catalyzed prepolymer-impregnated chopped strand mat are applied. Each layer is compacted by the use of brushes, squeegees, and/or rollers. The partially cured composite is then bag-molded by covering it with a thin sheet of silicone rubber and applying gas pressure while it is heated at a temperature of 177°C (350°F).

4.12 Spray-Up

In a slightly more sophisticated fabrication technique than hand layup, chopped strand and catalyzed prepolymer are deposited simultaneously onto an inexpensive mold surface. The buildup and curing process is similar to, but less labor-intensive than, the hand lay-up technique. The spray-up technique is sketched in Fig. 4.2.

4.13 Filament Winding

Most of the fabrication and testing techniques used by the polymer industry are adaptations of those developed by the metals and ceramics industries. However, filament-winding and pultrusion techniques are

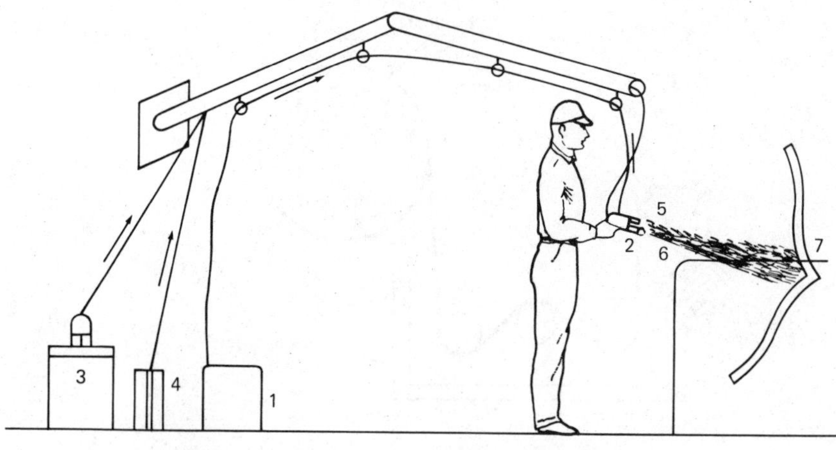

Spray-up process

Figure 4.2

unique fabrication processes developed and used by the reinforced-plastics industry.

In its simplest form, the filament-winding process consists of winding prepolymer-impregnated filaments, under tension, on a lathelike rotating cylinder, and this process is continued until the desired thickness is obtained. This cylindrical composite is then cured, usually in an oven, and the mandrel is removed.

The basic process has been improved to include conical shapes, cups, spheres, and other intricate shapes. The "wrap angles" may also be varied from low to high angles in successive layers.

Strong composite pipe is produced by using metal or thermoplastic pipe as the mandrel in the filament-winding process. Large rocket motor cases, high-pressure gas cylinders, golf clubs, automotive drive shafts, leaf springs, helicopter blades, and furniture are produced by the filament-winding process. This process is sketched in Fig. 4.3.

4.14 Pultrusion

Strong reinforced-plastic profiles are made by the pultrusion process, in which bundles of prepolymer-impregnated filaments are passed through a heated die of desired shape and the cure of the extrudate-like, partially cured profile is completed by heating outside the die. The basic pultrusion process has been upgraded by orientation of the

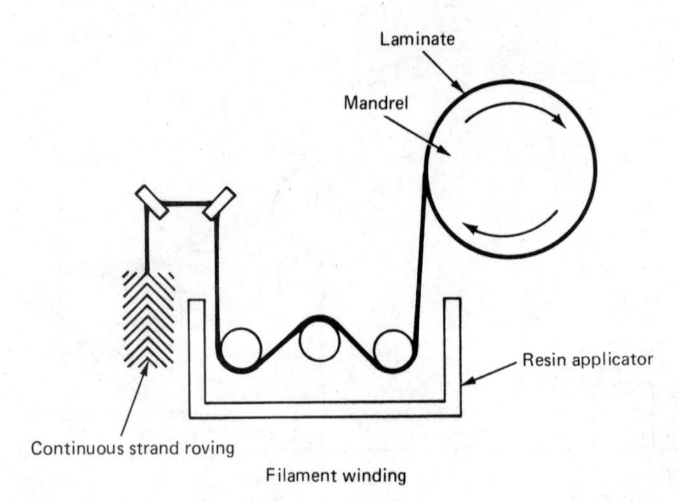

Figure 4.3

filaments and by the production of curved shapes, large sheets, and intricate shapes. Pipe, square tubing, large building panels, ladders, handrails, structural beams, fishing rods, and poles are produced by the pultrusion process. This process is sketched in Fig. 4.4.

4.15 Applications

Unsaturated polyesters and epoxy resins are used interchangeably as adhesives, potting compounds, moldings, and composites. The epoxy resins are preferred for use as coatings, potting compounds, and adhesives. They are used in composites when the higher cost of this superior product can be justified. Acrylic resin and acrylic elastomer-modified epoxy resin have superior adhesion to glass and other

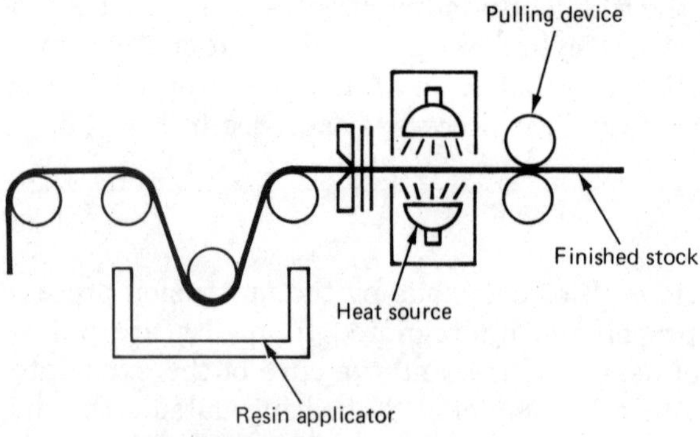

Figure 4.4 Continuous pultrusion.

surfaces. Because of their excellent properties, the use of epoxy resin composites, reinforced by graphite and graphite-aramide (hybrid)-reinforced composites will continue to increase.

4.16 Glossary

agent, linking A coupling agent.

amine, tertiary NR_3.

aramid An aromatic polyamide.

benzoyl peroxide An initiator of chain polymerization: $(C_6H_5CO_2)_2$.

bisphenol A $(HOC_6H_4)_2C(CH_3)_2$.

BMC Bulk-molding compound.

bulk molding The molding of a polyester compound (BMC) consisting of a log or rope of prepolymer, reinforcement, and other additives.

carbonization Pyrolysis.

Carothers' equation $\overline{DP} = 1/(1 - p)$ where p = extent of reaction.

chain reaction One which includes initiation, propagation, and termination steps.

cis Having functional groups on the same side of the double bond of an ethylene molecule.

coefficient of expansion The fractional change in dimensions of a test bar, per degree of temperature.

composite A substance consisting of a material (resin) in one or more other materials (reinforcements).

compressive strength The ability of material to resist a crushing force.

copolymer A polymer consisting of more than one repeating unit.

Corezin A vinyl ester.

Corrolite A vinyl ester.

cross-linked polymer A three-dimensional (network) thermoset polymer.

D Diameter.

Derakane A vinyl ester.

dielectric constant A measure of electric charge stored per unit volume between plates at unit potential.

dermatitis Irritation and possible blistering of skin.

DETA Diethylenetetramine, used as a curing agent for epoxy resins.

doctor's bar A device which controls the amount of prepolymer, etc., on a roller or spreader.

DMC Dough-molding compound.

DP Degree of polymerization; the number of repeating units in a polymer molecule.

drier A catalyst, such as cobalt naphthenate, which reduces surface stickiness of FRPs.

E Modulus.

E_{fu} The ultimate breaking strength of a fiber.

ejection The removal of a molded piece from the mold.

elongation at break The percent elongation at the moment of rupture of a test specimen.

epoxy group (—C—C—) with O bridging the two carbons

epoxy resin The polymer produced by the condensation of an aromatic diol, such as bisphenol A, and an epoxy (oxirane) compound, such as epichlorohydrin. The precursor may also be an epoxy compound produced by the reaction of cyclohexane and peracetic acid.

ethylene glycol $HOCH_2CH_2OH$.

exotherm The evolution of heat.

fiber A relatively short length of a filament, with a minimum length/diameter ratio of 100.

fiberglass Fiber made from glass filament produced by forcing molten glass through small holes (bushings).

filament A continuous fiber.

filament winding A fabrication process in which resin-impregnated filaments are wound on a rotating mandrel and cured.

flame retardant An additive which decreases the burning rate of a polymer.

flexural strength The resistance of a test specimen to bending stress.

flow The movement of a polymer, usually under pressure, at an elevated temperature.

free radical An electron-deficient molecule.

FRP Fiberglass-reinforced plastic.

fumaric acid The trans geometric isomer of *cis*-maleic acid.

gel coat The first layer of resin applied to the mold surface when making a plastic laminate.

graphitization The production of graphite by pyrolysis of carbon at elevated temperatures [> 1800°C (3272°F)] in an inert atmosphere.

hand lay-up The process in which successive layers of resin-impregnated fiberglass mat are placed on a mold by hand.

hardness Resistance to surface indentation.

heat deflection temperature DTUL; formerly called heat distortion temperature.

HET Chlorendic anhydride: $C_9H_2Cl_6O_3$ (hexacloroendomethylenetetrahydrophthalic anhydride).

Hooke's law Stress is proportional to strain.

hydrolysis Decomposition in the presence of water.

impact strength The ability of a test specimen to resist fracture when subjected to a hammerlike blow.

interface The junction between a resin and the fiber surface.

isophthalic acid *meta*-Phthalic acid.

Izod impact test The standard ASTM destructure impact test, in which the specimen usually includes a notch, with specified indentations.

Kelley-tyson One of the prevalent concepts of estimating stress buildup of reinforced fibers.

Kevlar An aramid, poly-*p*-terephthalamide.

L Load.

laminate A composite consisting of layers of resin and reinforcing sheets.

law of mixtures Properties of a mixture that are related to the relative amounts of additives present in the mixture.

L_c Critical fiber length.

Maleic anhydride

```
   HC=CH
   |   |
O=C     C=O
   \   /
    O
```

Mandrel The core around which resin-impregnated fiberglass is wound in fabrication processes, such as filament winding.

matched metal molding Compression molding of reinforced plastics.

matrix A resin.

modulus The ratio of stress to strain.

Mohs' hardness A measure of hardness using numbers from 1 to 10 of which the diamond has a hardness of 10.

mold The cavity in which the prepolymer is placed, where it assumes the shape of the cavity, usually at a high temperature and pressure.

monomer The building block of a polymer.

organosilane An R_4Si coupling agent.

organotitanate A coupling agent based on titanium compounds.

oxirane An epoxy group.

p Extent of a condensation reaction.

PAN Polyacrylonitrile.

parameter An arbitrary reproducible numerical value assigned to a specific property.

parting agent A mold release agent (wax).

pascal Pa, a unit of force; 1 $lb/in^2 = 6.895 \times 10^3$Pa.

pendant groups Functional groups on the polymer chain.

phase, continuous The resin matrix in a composite.

phase, discontinuous A reinforcing member of a composite.

phlegmatized paste A mixture of initiator (catalyst) and solvent or plasticizer.

plastic flow Deformation under load.

plastic tooling FRP tools used for metal forming.

Poisson ratio The ratio of the contracting strain to the elongation strain of a rod stretched at both ends.

polarity The charge on a molecule.

polyesters Polymers containing ester groups (—COOR).

promoter An accelerator or catalyst.

pultrusion A fabrication process in which a bundle or prepolymer-impregnated filaments are passed through a heated die and cured.

R· A free radical.

radical chain polymerization One in which the initiating species is a free radical (R·).

Rockwell hardness A number related to the depth of penetration of an indenter, under a specified load.

roving A collection of bundles of filaments.

RP Reinforced plastic.

shear Applied force which tends to cause one surface to pass by another as in the buttering of a piece of toast.

spinneret Tiny holes through which a resin is forced in the production of filaments.

spray-up A fabrication technique in which glass staple and resin are deposited simultaneously by use of a spray gun.

staple fiber A short-length filament.

strain The rate of change in length to the original length, i.e., the percent elongation.

strand A bundle of continuous filaments gathered together as a single unit.

strength, flexural The resistance of a test specimen to bending under load.

stress The rate of applied load to the cross-sectional area.

styrene $CH_2{=}CHC_6H_5$.

surfacing mat The mat of highly filamentized fiberglass used to produce a smooth surface as the first layer of FRP.

syntactic foam A plastic containing hollow beads.

tensile strength Resistance to a pulling force.

thermoplastic A polymer which can be reversibly heat-softened and hardened by cooling.

thermoset An infusible, insoluble polymer which cannot be softened by heat.

TMC Thick-molding compound.

trans Having functional groups on alternate sides of the double bond of an ethylene molecule.

transition temperature The temperature at which a change in properties occurs.

V Volume.

veil An ultrathin surface mat.

viscosity Resistance to flow.

vinyl ester An unsaturated ester, based on the condensation of methacrylic acid and the adduct obtained by the condensation of bisphenol A and propylene oxide.

Volan A methacrylatochromic chloride coupling agent.

wrap angle The angle at which filament is wound on the mandrel in the filament-winding process.

4.17 References

Ashton, J. E., J. C. Halpin, and P. H. Petit: *Pioneer Composite Materials,* Technomic Publishing Company, Stanford, Conn., 1969.

Beck, R. D.: *Plastic Product Design,* Reinhold Company, Inc., New York, 1980.

Bjorksten, J.: *Polyesters and Their Applications*, Van Nostrand-Reinhold Pub. Co., New York, 1956.

Bornig, H. V.: *Unsaturated Polyesters: Structure and Properties,* Elsevier Publishing Company, Amsterdam, Netherlands, 1964.

Burns, R.: *Polyester Molding Compounds,* Marcel Dekker, Inc., New York, 1982.

Dusek, K.: *Epoxy Resins and Composites,* Springer-Verlag, Heidelberg, N.Y., 1987.

Hill, D.: *An Introduction to Composite Materials,* Cambridge University Press, New York, 1981.

Lawrence, J. R.: *Polyester Resins,* Van Nostrand Reinhold Company, Inc., New York, 1975.

Mantell, C. L. (ed.): *Carbon and Graphite Handbook,* John Wiley & Sons, Inc., Interscience Publishers, New York, 1971.
May, C. A.: *Epoxy Resins,* Springer-Verlag New York Inc., New York, 1970.
McAdams, L. V., and J. A. Gannon: "Epoxy Resin," in J. I. Kroschwitz (ed.), *Encyclopedia of Polymer Science and Engineering,* vol. 6, John Wiley & Sons, Inc., Interscience Publishing, New York, 1986.
Nielsen, L. E.: *Mechanical Properties of Polymers and Composites,* Marcel Dekker, Inc., New York, 1971.
Potter, W. G.: *Epoxy Resins,* Springer-Verlag New York, New York, 1970.
Seymour, R. B. (ed.): *Additives for Plastics,* vols. 1 and 2, Academic Press, Inc., New York, 1978.
Tsai, S. W., and H. Hahn: *Introduction to Composite Materials,* Technomic Publishing Company, Westport, Conn., 1980.

Chapter

5

Inorganic Polymers

5.1 Glass

Glass is one of the oldest commercial polymers. The Egyptians produced useful articles from glass some 9000 years ago. One of the oldest known glass articles is a molded amulet of blue glass, which was cast in Egypt in 7000 B.C.

In the absence of a science to explain structure and property relations, the Egyptians, Assyrians, and Syrians developed a glass-based handicraft that was superior to that used for any other classic material of construction. Many of the advances made over a period of several centuries paralleled the more rapid advances that have been made in the plastics industry. The early artisans made ornaments, but the molding of glass became possible with the invention of the blowpipe.

Window glass, which was invented by a monk in the twelfth century, became widely used in the fifteenth century. In contrast to plastics, which some users have labeled "shoddy materials," glass was classified as a highly valued gem as late as the Ptolemaic era.

The making of sheet glass and bottles was automated in the early years of the twentieth century, but the making and use of glass remained an art for many decades.

Soda lime glass, which is the major glass, is produced by heating soda ash (Na_2CO_3), limestone ($CaCO_3$), sodium sulfate (Na_2SO_4), charcoal (C), and sand (SiO_2) in a furnace. The glass is shaped while it is above the T_g and then annealed in a heat chamber which is called the lehr. Internal strains in the glass are reduced during the annealing process.

Glass is a three-dimensional, amorphous polymer consisting of tetrahedrally linked hexagons containing siloxane $\leftarrow\!\!\mathrm{SiO}\!\!\rightarrow$ repeating units in the polymer chain. Like amorphous organic polymers, glass is transparent, has a high coefficient of expansion, has low thermal and electric conductivity, is thermoplastic, and exists as a high-viscosity liquid above its T_g. The coefficient of thermal expansion is reduced from 4×10^{-6} inches per inch per degree Fahrenheit [in/(in · °F)] for soda lime glass to 2×10^{-6}in/(in · °F) by the addition of borax ($Na_2B_4O_7$) to produce a borosilicate glass (Pyrex, Kimax).

As in the case of organic plastics, the art preceded the science of glass, and thousands of years elapsed before information on glassmaking was published by a priest named J. Neri, who discussed this art in the book *L'arte Vetraria,* which was published in Florence in 1612. Subsequently, an English physician discussed the science of glassmaking, and Kunckel added many concepts on glassmaking techniques.

In contrast to scientific investigations of organic plastics, which were shunned by the leading scientists of the late nineteenth and early twentieth centuries, extensive research on the relation of composition to optical properties were conducted by leading scientists such as Farady, Schott, Abbé, and Winkelman, and this research aided in the development of glass science and technology.

Silica crystallizes in rocks as quartz, tridymite, and cristobalite. Quartz, which is one of the most abundant minerals in the earth's crust, exists as a three-dimensional interconnected network of six-membered Si—O rings, i.e., three SiO_4 tetrahedra. Like some organic polymers, quartz is a piezoelectric substance, i.e., it generates electric polarization when subjected to external stress. Since only specific frequencies of electric current are able to pass through quartz crystal, these crystals are used in radio, television, radar, and watches.

Asbestos, which exists as a two-dimensional silicate sheet $[(Si_4O_{10})^{4-}]$, has been used for over 2000 years as a flame-resistant fiber. Asbestos, which was called "amiantus" by the Romans, exists as amphibole and serpentine asbestos. Chrysotile, which is a submember of the serpentine group, has the empirical formula $Mg_3Si_2O_5(OH)_4$ and is the most widely used type of asbestos.

Asbestos was used by Baekeland for reinforcing phenolic resins and has been widely used for reinforcing vinyl (PVC) floor tile, as well as for automotive brake linings.

However, when present in the air used for breathing, particles of asbestos cause asbestosis, bronchogenic cancer, and mesothelioma. Hence, the use of asbestos is being curtailed in applications in which it is not an essential ingredient.

Feldspar, granite, and sand also contain silica networks. Feldspar also contains aluminum silicate. Granite, which is derived from molten lava (rhyolite), consists of feldspar and quartz, and sand is a contaminated silica.

It is now recognized that alkali metals, such as sodium (Na), modify the polyhedra and triangular silica (SiO_2) networks. Boric (BO_3) triangles and germanic (GeO_2) tetrahedra are present in Pyrex and germanic glass, respectively. Since crystal formation in glass is undesirable from an optical and mechanical viewpoint, impurities are added to prevent devitrification. The addition of sodium oxide (Na_2O) reduces the solubility of glass. Empirical tests are used to evaluate these effects on the durability of glass. A polariscope is used to determine the presence of strains in glass. However, a small amount of birefringence is acceptable in optical glass.

Young's modulus, which is a function of temperature, is determined by a resonant-frequency method, in which the resonant frequency during the bending of a vibrating glass beam is measured. Hardness is measured by scratch-resistance on the Mohs scale, in which the diamond is the hardest (10), talc is the softest (1), and quartz has an intermediate value of 7.

Some consumers object to the blending of recycled plastics with virgin plastic materials, but it is standard practice in the glass industry to add from 5 to 40 percent group scrap glass (cullet) to the glass furnace. Also, the techniques for converting glass to useful products by drawing, pressing, casting, and blowing are used with appropriate modifications in plastic fabrication. Vycor, which has a melting point of 1499°C (2730°F), is produced by adding about 4 percent boric oxide to molten silica glass.

Tempered glass is produced by a special heat treatment, and optical glass fibers are coated with a highly reflective polymeric coating that guides the light from one end of the fiber to the other. Foamed glass is produced by adding a gaseous propellant, such as air, to molten glass.

5.2 Hydraulic Cement

Hydraulic cements have been used as materials of construction for over 2000 years. The Romans mixed lime ($CaCO_3$) with volcanic ash to produce a hydraulic cement. Since the ash was obtained from a suburb of Naples called Pozzuoli, the material was called Pozzolan cement.

In 1824, J. Aspidin patented a hydraulic cement produced by the calcination of argillaceous limestone. Since the product resembled the building stone obtained from the Isle of Portland, Aspidin called this product Portland cement. Pozzolan cement and mixtures of Pozzolan cement and Portland cement continue to be used, but the major hydraulic cements are produced by heating a mixture of clay and limestone in a continuous rotary kiln and crushing the cooled clinkers. The most common proportions of reactants is 3 parts of limestone and 1 part of clay. The product is primarily a dicalcium or tricalcium silicate.

It is customary to grind the clinker so that over 98 percent passes through a 200-mesh sieve. The term "specific surface," i.e., the surface area in square centimeters per gram, is used to quantify the fineness of the ground clinker. A small amount of calcium sulfate (gypsum or anhydrite: $CaSO_4$) is added during the grinding to control the setting time of the cement.

Air-entrainment agents, such as tallow, are added to Portland cement to increase its resistance to scaling and to protect against damage associated with alternate freezing and thawing. Calcium aluminate cement (Lumnite cement) is produced by fusing mixtures of lime and bauxite (Al_2O_3). This high-alumina cement is quick-setting and is more resistant to seawater than Portland cement.

White cement is produced from pure calcite limestone. Natural white limestone is mined in eastern Pennsylvania, but darker varieties may be treated with fluorspar and used in place of the natural white product.

The tensile strength of concrete can be increased dramatically by the use of steel bars or rods as reinforcements. J. L. Lambot and T. Hyatt are credited with the simultaneous and independent development of reinforced-concrete beams in the 1850s. Monier patented this improved material of construction for making flowerpots in 1867 and then enlarged his structures to include water tanks and bridges. Freyssinet introduced the concept of prestressed concrete in the 1920s.

As in other composites, the stress on the concrete (continuous phase) is transferred to the reinforcement (discontinuous phase). Accordingly, it is customary to use bars with raised surfaces (deformed bars) to improve the mechanical bond between the two phases. Concrete is produced by the addition of coarse and fine aggregates to the cement mix.

Because of its low tensile strength, the use of concrete was limited prior to the development of reinforcing techniques. Fortunately, the coefficients of expansion of steel and concrete are similar. Otherwise, these structures would shatter when the temperature was increased or decreased.

5.3 Silicones

Polysilicic acid, $+Si(OH)_2-O+$ is a polysiloxane which is produced by the hydrolysis of silicon tetrachloride ($SiCl_4$). The hydrochloric acid produced must be removed, as formed, by dialysis.

Organic-substituted polysiloxanes (silicones) may be considered to be organic glass. These siloxanes $+SiO+$ have two alkyl (R) pendant groups on each silicon (Si) atom, and thus the surface of these polymer chains is similar to that of polyolefins. However, the bond energy of the siloxane backbone is higher than that of carbon-carbon backbones, and hence silicones can be used at relatively high temperatures. Since silicones have excellent resistance to moisture and excellent lubricity, they are used as water repellents and as lubricants.

Silicones were produced in the early 1900s by Kipping, who did not visualize any commercial utility for these products. However, chemists at General Electric and Dow Corning were able to develop the commercial products that are available today. As shown by Eq. (5.1), linear silicones are obtained by the hydrolysis of dialkyl dichlorosilanes (R_2SiCl_2). The propagating chains are terminated by the addition of small amounts of trialkylchlorosilane (R_3SiCl), and network polymers are produced by the incorporation of small amounts of monoalkyltrichlorosilanes ($RSiCl_3$).

$$R_2SiCl_2 + 2H_2O \xrightarrow{-2HCl} [R_2Si(OH)_2] \xrightarrow{-H_2O} +\!\!(SiR_2O)\!\!+ \qquad (5.1)$$

Silicones are essentially transparent to ultraviolet radiation and hence have good weatherability. Silicones retain their flexibility over a temperature range of −40°C (−40°F) to 249°C (480°F). Since there is little opportunity for interaction among the pendant alkyl groups, silicones have high torsional mobility and high permeability to gasses, such as oxygen.

Bouncing putty (Silly Putty) is produced by heating uncapped silicones with boric acid (H_3BO_3). As shown by Eq. (5.2), the borate end groups are present in the reaction product.

$$Cl \leftarrow\!\!(SiR_2O)\!\!\rightarrow_n SiR_2Cl \xrightarrow{(HO)_3B} (HO)_2B \leftarrow\!\!(SiR_2O)\!\!\rightarrow_n SiR_2B(OH)_2 \qquad (5.2)$$

The solubility of silicones in organic solvents is reduced when one of the methyl pendant groups in the repeating unit is replaced by a trifluoropropyl group.

Room temperature–vulcanizing (RTV) silicone sealants are available as one-and two-package materials. The one-package sealants have residual alkoxy (OR) or acetoxy groups which react with moisture in the air to form stable polymers.

The two-package silicone consists of prepolymers in one package and a catalyst [stannous octoate: $SnOOC(CH_2)_6CH_3$], cross-linking agents, and filler. The properties of silicones are shown in Table 5.1.

Liquid silicones (oligomers) are used in hydraulic, dielectric, damping, and power transmission, and for heat transfer. The higher-molecular-weight silicones are used as sealants, gaskets, electrical insulation, and caulking materials, and for mechanical parts.

The methylsilicones are biologically inert and are used as prosthetics. However, when heated above 279°C (535°F), trifluoropropyl silicones emit toxic compounds. The chlorosilanes react with water and emit hydrochloric acid, which is corrosive.

TABLE 5.1 Properties of Typical Silicones

Property, units*	RTV	Fiberglass-filled silicone	Graphite-filled silicone
Melting point T_m, °F	—	—	—
Glass transition temp. T_g, °F	—	—	—
Processing temp., °F	—	300	300
Molding pressure, 10^3 lb/in^2	--	1	1
Mold shrinkage, 10^{-3} in/in	5	5	5
Heat deflection temp. under flexural load of 264 lb/in^2, °F	--	500+	480
Maximum resistance to continuous heat, °F	250	500	450
Coefficient of linear expansion, 10^{-6} in/(in · °F)	250	15	20
Compressive strength, 10^3 lb/in^2	—	12	9
Izod impact strength, ft · lb/in of notch	—	1.5	4
Tensile strength, 10^3 lb/in^2	1	5	4
Flexural strength, 10^3 lb/in^2	—	10	9
Percent elongation	—	4	3
Tensile modulus, 10^3 lb/in^2	—	—	—
Flexural modulus, 10^3 lb/in^2	—	20	30
Rockwell hardness	M80	M90	M80
Specific gravity	1.5	2.0	—
Percent water absorption	0.1	0.2	—
Dielectric constant	4	4	—
Dielectric strength, V/mil	500	350	150
Resistance to chemicals at 75°F:†			
Nonoxidizing acids (20% H_2SO_4)	Q	Q	Q
Oxidizing acids (10% HNO_3)	U	U	U
Aqueous salt solutions (NaCl)	S	S	S
Polar solvents (C_2H_5OH)	S	S	S
Nonpolar solvents (C_6H_6)	Q	Q	Q
Water	S	S	S
Aqueous alkaline solutions (NaOH)	S	S	S

*lb/(in^2 · 0.145) = KPa (kilopascals); ft · lb/(in of notch · 0.0187) = cm · N/cm of notch.
†S = satisfactory; Q = questionable; U = unsatisfactory.

5.4 Polyphosphazenes

Polyphosphazenes [$\leftarrow PCl_2{=}N \rightarrow$] are produced by the thermal [249°C (480°F)] polymerization of the cyclic trimer $(PCl_2{=}N)_3$. Since this inorganic elastomeric polymer is unstable in the presence of moisture, the chlorine pendant groups are replaced by alkoxide groups (OR) by heating the linear white phosphonitrilic chloride polymer with sodium alkoxide (NaOR) in a solution of benzene and tetrahydrofuran.

Stable polymers have been produced by using sodium ethoxide ($NaOC_2H_5$), sodium trifluoroethoxide ($NaOCH_2CF_3$), or sodium phenoxide ($NaOC_6H_5$). Stable polymers have also been produced by reacting the polyphosphonitrilic chloride with amines, such as butylamine ($C_4H_9NH_2$) or aniline ($C_6H_5NH_2$).

These flexible, high-molecular-weight (2×10^6) polymers with alkoxide pendant groups are characterized by low T_g's [−66°C (−87°F)] and retain their flexibility at temperatures as high as 241°C (465°F). The opacity of these crystalline polymers can be reduced by copolymerization. For example, the copolymer of trifluoroethoxyphosphazene and octafluoropentoxyphosphazene is a clear amorphous polymer [T_g = −68°C (−90°F)]. It is customary to include an unsaturated monomer in order to produce a cross-linkable terpolymer which can be cured by the addition of an organic peroxide.

Polyphosphazenes are degraded at temperatures above 204°C (400°F). However, they are thermally stable and flame-resistant at lower temperatures and are used as gaskets, O-rings, fuel lines, and hoses. The clear phosphazene $\!+\!P(NCH_3)_2\!=\!N\!+\!$, produced by the reaction of polydichlorophosphazene with dimethylamine, is soluble in water.

5.5 Polysulfur Nitride

Polysulfur nitride [polythiazyl: $(SN)_n$] is produced by the solid-state polymerization of a cyclic dimer [$(SN)_2$], which is obtained by the reaction of ammonia (NH_3) and sulfur tetrafluoride (SF_4). Films of this crystalline inorganic polymer are produced by depolymerization of the polymer at 143°C (290°F), followed by repolymerization on an appropriate surface, such as PTFE (Teflon).

Polysulfur nitride is an electric conductor because of the cis-trans arrangement of the polymer chains. Brominated polysulfur nitride [$(SNBr_{0.4})_n$], which is produced by the reaction of bromine with $(SN)_n$, is also a conductor of electricity.

5.6 Other Inorganic Polymers

Several nonmetallic elements, such as antimony, arsenic, bismuth, nitrogen, phosphorus, and sulfur, tend to form low-molecular-weight polymers (oligomers), but only those elements in the carbon group, such as carbon, silicon, germanium, tin, and lead, tend to form moderately high molecular weight polymers. This tendency decreases with the decline in covalent bond energies as one goes from carbon to the elements of higher molecular weight. Hence, while silanes [$(Si)_n$], germanes [$(Ge)_n$], stannanes [$(Sn)_n$], and plumbanes [$(Pb)_n$] have been prepared, the only commercial organic homopolymers are those containing the carbon atoms.

5.7 Glass Structures

If one eliminates the fiberglass used as a reinforcement for polymeric composites, the principal use of glass as a structural material is in the form of sheets of specified thickness. Flat glass (float glass) is produced by flowing molten glass over molten tin.

Single-strength glass (glazing) transmits about 85 percent of visible light. It is notch-sensitive, and large cracks will propagate from small cracks or other imperfections present in this brittle transparent product. Tempered glass is produced by reheating annealed glass at 649°C (1200°F) and cooling the surfaces more rapidly than the interior. This glass is more resistant to bending, impact, and thermal stress than ordinary glass. However, any cutting or drilling of this glass must be done before tempering to prevent stress release and shattering of the glass. The strength of heat-strengthened glass is between that of annealed and that of tempered glass.

Shattering of glass can be prevented by laminating two sheets of annealed glass with a sandwich layer of polyvinyl butyral (PVB; Saflex). Multiple layers of laminated glass are used for bullet-resistant glass. PVB is produced by the condensation of butyraldehyde [$H(CH_2)_3CHO$] with polyvinyl alcohol [$\leftarrow CH_2CHOH \rightarrow$; PVAL]. PVAL is produced by the hydrolysis of polyvinyl acetate, $+CH_2CH(OOCCH_3)_n+$.

Clear silicone sealants are used to butt-joint sheet glass. Silicone, polyisobutylene, acrylics, and polyethylene sulfide (Thiokol LP-2) sealants are used to hold sheet glass in frames. Tapes of these materials and synthetic elastomers are also used as lock-strip gaskets.

Because the art of using thermoplastic sheets is much newer than that of glazing with plastic glass, little information is available on the design and use of plastic sheets. However, the inorganic and organic sheets have many common characteristics which permit some transfer of design data. More important, the organic sheet structures are more versatile, and this justifies their use where they exhibit superior functionality.

5.8 Concrete Structure

Concrete made from Portland cement (25%) and graded aggregate (75%) has excellent compressive strength (4000 lb/in^2). Since the hydration of tricalcium silicate and aluminate is a slow process, 7 days is usually required for the room-temperature cure of Portland cement.

Concrete blocks contain sand and gravel aggregates or graded sand. Asbestos-reinforced cement slabs are called Careystone. Aluminum oxide grains are sprinkled on partially set cement to produce nonslip concrete. When aluminum powder is added to the mortar, a porous structure (Aerocrete) is obtained as a result of the formation of a hydrogen propellant produced by the reaction of alkali and aluminum.

Another low-density cement (Haylight, Superock) is produced by adding steam to the molten slag. Porous aggregate is also added as a filler to produce low-density concrete (Flufrok). Conductive hydraulic cement is produced by the addition of carbon black. The plastics industry has adopted many of these techniques for the production of strong plastic concrete.

Over 900,000,000 t of hydraulic cement is consumed annually, worldwide.

5.9 Glossary

Abbé E. A scientist who investigated the relation of optical qualities of glass to composition. An instrument used to measure the index of refraction is named after him.

Aerocrete Low-density concrete.

alkoxide group —OR.

alkyl group $H(CH_2)_n$—.

aluminum Al.

amiantus The Roman name for asbestos.

amine RNH_2; an organic derivative of ammonia (NH_3).

amorphous Noncrystalline.

amphibole A form of asbestos.

anhydrite $CaSO_4$.

annealing Removal of internal strain by thermal treatment.

antimony Sb.

arsenic As.

asbestos $Mg_3Si_2(OH)_4O_5$, a naturally occurring fibrous magnesium silicate.

asbestosis A chronic lung inflammation resulting from inhalation of asbestos particles.

Aspidin, J. The first patentee of a hydraulic cement.

Baekeland, L. The inventor of PF plastics (Bakelite).

bauxite Al_2O_3.

birefringence The splitting of a beam of light into two components.

bismuth Bi.

boria BO_3.

carbon C.

carbon black An amorphous form of carbon, produced by the incomplete combustion of hydrocarbons.

Careystone An asbestos-reinforced concrete.

cement, hydraulic Cement which hardens when water is added.

cement, Lumnite Calcium aluminate cement.

cement, Portland The most common hydraulic cement.

cement, Pozzolan A hydraulic cement produced by mixing burnt clay and volcanic dust.

cement, reinforced A hydraulic cement reinforced by steel bars.

chrysotile $Mg_3Si_2(OH)_4O_5$: serpentine asbestos.

clay A finely crystalline hydrous silicate.

clinker Vitrified material from a furnace.

coefficient of expansion The fractional change in length per degree Fahrenheit.

cristobalite A form of asbestos.

cullet Waste or scrap glass, usually ground.

devitrification The process of changing from an amorphous to a crystalline form.

dialysis Selective diffusion through a membrane.

energy, bond The energy required to cleave the bond between two atoms.

feldspar A silicate mineral that makes up 60 percent of the earth's crust.

fiberglass Filaments formed by passing molten glass through small holes (spinnerets).

Flufrok A low-density concrete.

fluorospar A (fluorite) CaF_2 mineral with a Mohs' hardness of 4.

germania GeO_2.

glass A hard amorphous transparent, brittle material produced by the fusion of silicates with basic compounds, such as sodium carbonate (Na_2CO_3) and/or limestone ($CaCO_3$).

glass, float Sheet glass cast on molten tin.

glass transition temperature T_g The temperature at which a brittle amorphous, glasslike solid becomes somewhat flexible as the temperature is increased.

glazing Fitting a pane of glass into a frame.

granite Crystalline plutonic rock composed of quartz and feldspar.

group, pendant A group on the main polymer chain.

gypsum $CaSO_4$.

Haylight A low-density concrete.

Kimax A borosilicate glass.

Kipping, F. The inventor of silicones.

Kunckel An inventor who improved glass making.

laminate A sheet consisting of several layers cemented together.

lead Pb.

lehr A heat chamber used for annealing glass.

lime CaO.

limestone $CaCO_3$.

LP-2 A liquid polyethylene sulfide (Thiokol) which can be converted to a solid in the presence of lead oxide.

lubricity Slipperiness.

mesothelioma A chronic inflammation of the epithelium lining in body cavities.

Mohs' scale A scale from 1 to 10 based on scratch hardness. Diamond = 10; talc = 1.

monomer, unsaturated A building block with a vinyl group +CH=CH+.

Neri One of the first to publish information on glassmaking.

network polymer A three-dimensional (cross-linked) polymer.

oligomer A low-molecular-weight polymer.

peroxide, organic ROOR.

phase, continuous The resinous or nonfiller component of a composite.

phase, discontinuous The filler or reinforcement component of a composite.

phosphorus P.

piezoelectric effect A generation of electrical polarization when subjected to stress.

polariscope An instrument for determining the effect of polarized light as it passes through the material being tested.

polyphosphazene $\text{+}PCl_2\text{=N+}$.

polysilicic acid $\text{+Si(OH)}_2\text{—O+}$.

polysulfur nitride $(SN)_n$.

polyvinyl acetate PVAC: $-\!\!(CH_2CH(OOCCH_3)\!)\!\!-$.

polyvinyl alcohol PVAl: $[\!-\!(CH_2CHOH)\!-\!]$; produced by the hydrolysis of polyvinyl acetate.

polyvinyl butyral PVB, a clear polymer derivative produced by the condensation of butyraldehyde with polyvinyl alcohol.

prosthetic device An artificial replacement for a body part.

putty, bouncing Silicone with borate terminal groups.

Pyrex A borosilicate glass.

quartz Natural crystals of SiO_2, with a Mohs' hardness of 7 and a specific gravity of 2.65.

R An alkyl group.

rhyolite Molten lava.

Saflex A trade name for two sheets of glass with an inner layer of polyvinyl butyral.

serpentine A type of asbestos.

silica SiO_2.

silicate Compounds containing SiO_4 tetrahedra.

silicon Si.

silicone A polysiloxane with alkyl pendant groups: $-\!(Si(R_2)—O)\!-$.

siloxane $-\!(Si—O—Si—O)\!-$.

specific surface The surface area in square centimeters per gram.

stannous octoate $SnOOC(CH_2)_6CH_3$; the tin salt of octoic acid.

strength, tensile The ability to withstand a stretching load.

sulfur S.

tallow Animal fat.

tridymite SiO_2, a polymorph of quartz.

trifluoropropyl group $F_3C(CH_2)_2—$.

vinyl A commonly used name for PVC.

viscosity Resistance to flow.

Vycor A high-melting-point glass.

Young's modulus The ratio of tension stress to strain.

5.10 References

Allcock, H. R.: *Phosphorus-Nitrogen Compounds,* Academic Press, Inc., New York, 1972.

Allen, E.: *The Professional Handbook of Building Construction,* John Wiley & Sons, Inc., New York, 1985.
Bogue, R. H.: *The Chemistry of Portland Cement,* Reinhold Publishing Corporation, New York, 1947.
Lea, F. M.: *The Chemistry of Cement and Concrete,* 3d ed., Edward Arnold (Publishers) Ltd, London, 1971.
Phillips, C. J.: *Glass, the Miracle Maker, Its History, Technology, Manufacture,* Pitman Publishing Corporation, New York, 1948.
Powers, T. C.: *The Properties of Wet Concrete,* John Wiley & Sons, Inc., New York, 1968.
Rochow, E. G.: *Chemistry of Silicones,* John Wiley & Sons, Inc., New York, 1946.
Stone, F. G. A., and W. A. G. Graham (eds.): *Inorganic Polymers,* Academic Press, Inc., New York, 1962.

Chapter

6

Elastomers for Engineering Applications

6.1 Introduction

While the major use of elastomers will continue to be for tire construction, these unique materials are also used and will continue to be used in many other engineering applications. Over 14,000,000 t of synthetic elastomers (SR) and 3,000,000 t of NR were used worldwide in 1987. The consumption of these elastomers in the United States in 1987 was 2,000,000 and 750,000 t, respectively. The U.S. production of SR in 1987 was as follows: 800,000 t of SBR, 360,000 t of polybutadiene rubber (BR), 180,000 t of ethylene-propylene rubber (EPDM), 60,000 t of NBR, 90,000 of neoprene (CR), and 135,000 t of butyl rubber (IIR). In 1987 50,000 t of CR was exported from the United States.

6.2 Characteristics of Elastomers

An elastomer, which is a general term for natural and synthetic rubbers, possesses unique properties which are not characteristic of any other material of construction. Intermolecular forces between the chains of these macromolecules are weak London dispersion forces, which are less than 20 percent of the strong intermolecular forces present in fibers. Since these forces are weak, these materials are soft and sticky and tend to cold flow, i.e., the polymer chains tend to move slowly against each other with a minimal restriction to such movement. However, as demonstrated accidentally by Charles Goodyear in 1838, this tendency to change shape in the absence of strong external stress can be reduced by the introduction of an occasional cross-link between the chains.

Since there is a distance of at least 100 atoms between these rela-

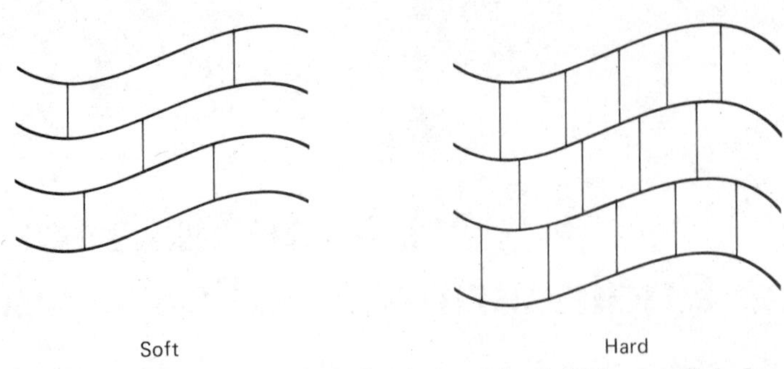

Figure 6.1 Soft (low cross-link density) and hard (high cross-link density) rubbers.

tively strong cross-links, the non-cross-linked sections of the chains (principal sections) are able to exhibit elastic properties in the absence of any cold flow. These soft elastomers, with a small number of cross-links (long principal sections) are said to have a low cross-link density, in contrast to highly cross-linked macromolecules, like hard rubber (ebonite), which have a high cross-link density, i.e., very short principal sections. The differences in molecular structure between soft and hard rubbers are shown by the simulated chains in Fig. 6.1.

Since a small amount of cross-linking is essential for the prevention of cold flow, the chains of elastomeric molecules are usually unsaturated; i.e., they contain ethylenic double bonds which are capable of forming cross-links. Thus, the building blocks of NR and of most SRs are dienes. The building block for NR is 2-methylbutadiene [$H_2C{=}C(CH_3){-}CH{=}CH_2$], and the repeating unit in the polyisoprene chain is $+CH_2C(CH_3){=}CHCH_2+$.

The unstretched molecules of an elastomer are amorphous, and the polymer chains are coiled like pieces of cooked spaghetti. Entropy S is a measure of disorder, and hence unstretched rubber has a high entropy value. However, this stretching of the polymer chains is reversible, and since the disorderly (high entropy) state is preferred, the stretched polymer chains return to their original coiled conformations when the stress is removed.

If one places an unstretched rubber band against the lips, one will observe that the band becomes warm when stretched. Since this effect was first recorded by a blind scientist named Gough and quantified by Joule almost two centuries ago, it is called the Gough-Joule effect. This effect may be demonstrated simply and dramatically by showing that a stretched rubber band contracts when heated.

One may also observe that an unfilled rubber band becomes opaque when stretched, i.e., the amorphous, randomly coiled chains are aligned and form crystallites when stretched. These crystalline do-

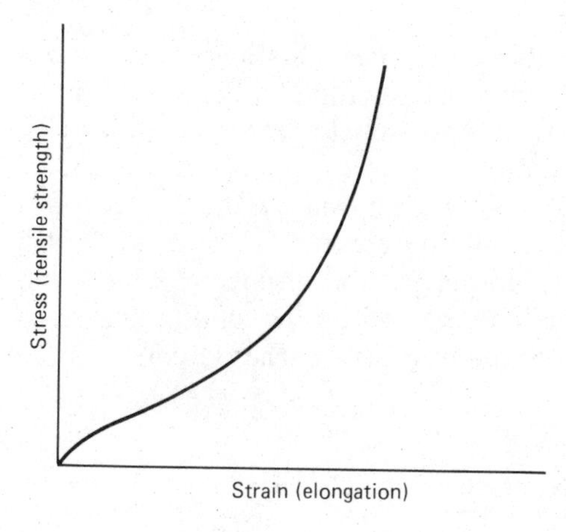

Figure 6.2 Stress-strain curve of an elastomer.

mains (crystallites) act as physical cross-links between chains and thus reinforce the macromolecules so that they exhibit high modulus (stiffness).

This effect is more pronounced when carbon black is present as a reinforcing filler. Elastomers such as SBR, which have less tendency to crystallize than NR, are not particularly useful as structural elastomers unless they are filled with carbon black. The surface of carbon black particles is attracted to the elastomeric polymers at their interfaces so that these points of attraction are actually physical cross-links between polymer chains.

The lengthwise extension of a stretched elastomer is accompanied by a transverse contraction. This deformation at constant volume for an ideal elastomer (Poisson ratio, ν) is 0.5. For glassy polymers ν is equal to about 0.33, and for elastomers it is equal to about 0.5.

The stress-strain curve for a hookean solid is a straight line. However, as shown in Fig. 6.2, the curve for an elastomer deviates from the straight line because of the high elongation, which is related to chain disentanglements, and approaches a straight line after a large extension (elongation of 500%) because of an increase in intermolecular forces of the aligned low-entropy chains.

The tensile strength of vulcanized NR is about 1000 lb/in^2, and its elongation, before rupture, is about 1000 percent. In contrast, hard rubber has a tensile strength of about 8500 lb/in^2 and an elongation of about 1 percent.

6.3 Neoprene

CR, which is a polymer of chloroprene (2-chlorobutadiene: $H_2C{=}CClCH{=}CH_2$), was one of the first synthetic elastomers and is still used today as a heat- and solvent-resistant elastomer. CR retains its elastomeric properties at temperatures as high as 121°C (250°F), it is less permeable to gases and more resistant to ozone (O_3) than NR, and it does not support combustion.

CR [$\leftarrow CH_2—CCl = CH — CH_2 \rightarrow$] is produced by the free-radical-initiated emulsion polymerization of chloroprene. In contrast to NR, which is a *cis*-poly-2-methyl-1,3-butadiene, neoprene is almost entirely a *trans*-poly-2-chlorobutadiene, i.e., while the chain extensions in NR are all on the same side of the ethylenic carbon atoms

$$(—\overset{|}{C} = \overset{|}{C}—)$$

they are on opposite sides in neoprene

$$(—\overset{|}{C} = \underset{|}{C}—)$$

Thus, neoprene crystallizes readily when stretched and is said to exhibit high gum tensile strength. Also, in contrast to NR, which is vulcanized (cured) by heating with sulfur (S), CR is cured by heating with zinc oxide (ZnO) or magnesia (MgO). These agents form cross-links by reacting with the chlorine pendant groups in the polymeric chains of neoprene.

Solutions of neoprene are used as adhesives and coatings, and molded neoprene is used as gaskets, V-belts, mine conveyor belts, motor mounts, and highway joint seals. Neoprene is also extruded as hose and as wire and cable coatings. While the monomer chloroprene is a toxic chemical, there is little if any monomer in the molded or extruded neoprene articles.

The properties of neoprene are shown in Table 6.1.

6.4 Polybutadienes

Since vulcanized NR (*Hevea braziliensis; cis*-poly-2-methyl-1,3-butadiene) was used for almost a century before synthetic elastomers were commercially available, considerable useful engineering data were developed, and this information is applicable to synthetic poly-1,3-butadiene (BR) and poly-2-methylbutadiene (IR). Four different isomeric polymers can be produced by the polymerization of 2-methyl-1,3-butadiene (isoprene), but a 99 percent yield of *cis*-poly-1,4-isoprene is obtained by the use of trimethylaluminum-titanium tetrachloride catalyst in a hydrocarbon solvent. Both the natural and synthetic *cis*-polybutadienes have low T_g values (−73°C; −100°F) and are used to make radial tires.

Expoxidized NR, which is produced by heating natural rubber with peracetic acid and copolymers of NR and methyl methacrylate (Heveaplus) are available commercially. The techniques used to make

TABLE 6.1 Properties of Neoprene (CR)

Property, units*	Vulcanized unfilled CR	Carbon-black-reinforced CR
Melting point T_m, °F	—	—
Glass transition temp. T_g, °F	49	52
Processing temp., °F	—	—
Molding pressure, 10^3 lb/in^2	—	—
Mold shrinkage, 10^{-3} in/in	—	—
Heat deflection temp. under flexural load of 264 lb/in^2, °F	—	—
Maximum resistance to continuous heat, °F	150	160
Coefficient of linear expansion, 10^{-6} in/(in · °F)	—	—
Compressive strength, 10^3 lb/in^2	—	—
Izod impact strength, ft · lb/in of notch	—	—
Tensile strength, 10^3 lb/in^2	4.5	3.5
Flexural strength, 10^3 lb/in^2	—	—
Percent elongation	900	550
Tensile modulus, 10^3 lb/in^2	0.25	0.6
Flexural modulus, 10^3 lb/in^2	60	75
Shore A hardness	1.2	114
Specific gravity	5	5
Percent water absorption	2	2
Dielectric constant	750	750
Dielectric strength, V/mil	—	—
Resistance to chemicals at 75°F:†		
Nonoxidizing acids (20% H_2SO_4)	S	S
Oxidizing acids (10% HNO_3)	Q	Q
Aqueous salt solutions (NaCl)	S	S
Polar solvents (C_2H_5OH)	S	S
Nonpolar solvents (C_6H_6)	Q	Q
Water	S	S
Aqueous alkaline solutions (NaOH)	S	S

*lb/(in^2 · 0.145) = KPa (kilopascals); ft · lb/(in of notch · 0.0187) = cm · N/cm of notch.
†S = satisfactory; Q = questionable; U = unsatisfactory.

synthetic *cis*-polybutadiene have been used to produce copolymers and blends which should have superior qualities to NR for engineering applications. Time-tested design and performance data on NR tires and mechanical goods can be used for other applications of NR and polybutadienes.

Butadiene and isoprene are considered to be toxic chemicals but neither of these monomers is present in vulcanized polybutadienes.

The properties of typical polybutadienes are shown in Table 6.2.

6.5 Polysulfide Elastomers

Polysulfide elastomers (Thiokol), which are produced by the condensation of sodium polysulfide (Na_2S_x) and an organic dichloride, such as ethylene dichloride [$Cl(CH_2)_2Cl$], are amorphous polymers which do

TABLE 6.2 Properties of Typical Carbon-Black-Reinforced, Vulcanized Polybutadienes

Property, units*	NR	IR
Tensile strength, 10^3 lb/in^2	4	3
Percent elongation	500	500
300% modulus, 10^3 lb/in^2	2	1.2
Shore A hardness	62	62
Specific gravity (gum stock)	0.9	0.9
T_g (gum stock), °F	−98	−98
Resistance to chemicals at 75°F:†		
Nonoxidizing acids (20% H_2SO_4)	S	S
Oxidizing acids (10% HNO_3)	U	U
Aqueous salt solutions (NaCl)	S	S
Aqueous alkaline solutions (NaOH)	S	S
Polar solvents (C_2H_5OH)	S	S
Nonpolar solvents (C_6H_6)	U	U

*lb/(in^2 · 0.145) = KPa (kilopascals).
†S = satisfactory; Q = questionable; U = unsatisfactory.

not crystallize when stretched, and hence reinforcing fillers, such as carbon black, must be added to obtain relatively high tensile strengths. Thiokol may be vulcanized in the presence of zinc oxide (ZnO) and thiuram accelerators, such as tetramethylthiuram disulfide (Tuads). The accelerators modify the sulfur links and serve as chemical plasticizers.

A typical Thiokol with 60 parts of carbon black per 100 of polymer has a tensile strength of 1200 lb/in^2, an elongation of 380 percent, a specific gravity of 1.25, and a Shore A hardness of 68. Vulcanized Thiokol ST has good resistance to compression at a temperature range of −40°C (−40°F) to 66°C (150°F). It has excellent resistance to ozone (O_3) and ultraviolet radiation. Thiokol has a low permeability to solvents, such as gasoline; esters, such as ethyl acetate; and ketones, such as acetone.

The principal application of solid Thiokol elastomers is as rollers, which are used for lacquering cans, hose lines, and gas-meter diaphragms. Thiokol sealants are produced by reducing the solid polymer and then converting the liquid polysulfide (LP-2, LP-3) to a solid by oxidation with lead oxide. These sealants are widely used as binders for solid propellants and as caulking materials.

6.6 Silicone Elastomers

Silicone elastomers (SIs) are useful over a wide range of temperature [−101°C (−150°F) to 316°C (600°F)]. There is little change in Young's modulus as the temperature is lowered [10,000 lb/in^2 at −101°C (−150°F)], but the relatively low tensile strength is satisfactory at high temperatures and is comparable to that of NR at 204°C (400°F). The low-temperature flexibility of polydimethylsiloxane is improved by replacing some of the methyl groups (CH_3) with phenyl groups (C_6H_5) or ethyl groups (C_2H_5).

It is customary to replace some of the methyl groups with vinyl groups ($H_2C{=}CH$—) in polydimethylsiloxane in order to produce a vulcanizable SI. Most SIs are reinforced by fumed silica. Carbon black provides moderate reinforcement but interferes with the peroxy curing (vulcanization).

Flexible articles such as gaskets may be fabricated by compression or injection molding, and tubing is produced by extrusion at moderate temperature. The formed objects are cured by steam after fabrication.

A typical solvent-resistant silicone has a tensile strength of 1000 lb/in^2, an elongation of 700 percent, a Shore A hardness of 70, a brittle point of −68°C (−90°F), and a dielectric constant of 3.

6.7 Acrylic Elastomers

Elastomeric copolymers of 2-chloroethyl acrylate and vinyl isobutyl ether (Lactoprene; ACM) were developed by C. Fisher in the late 1950s. These solvent-resistant elastomers, which have a tensile strength of 2100 lb/in^2 at an elongation of 200 percent and a Shore A hardness of 80, are used as O-rings, oil seals, and gaskets. ACM has excellent resistance to ozone, hot oils, ultraviolet radiation, and mineral oils.

6.8 Ethylene-Propylene Elastomers

While polyethylene and isotactic polypropylene (it PP) are hard crystalline polymers, the random copolymer (EPM) is an elastomer, which can be cured by the addition of organic peroxy compounds. However, the terpolymer (EPDM), which includes a diene (D) monomer, such as norbornene, is readily cured by heating with sulfur. EPDM is resistant to ozone and ultraviolet radiation and can be extended by the addition of mineral oils.

EPDM has excellent resistance to acids, alkalies, and aqueous salt solutions but has limited resistance to hydrocarbon solvents. This colorless elastomer is used as a cable coating, hose, and single-ply roofing (SPR). The technique used for EPDM-SPR can be adapted to many other engineering applications. The properties of a typical vulcanized EPDM elastomer are shown in Table 6.3.

6.9 IIR

IIR is a copolymer of isobutylene [$CH_2{=}C(CH_3)_2$] with a small amount of isoprene (5%). Thus, there is enough unsaturation in IIR to

TABLE 6.3 Properties of a Typical Vulcanized Reinforced EPDM Elastomer

Property, units*	
Tensile strength, 10^3 lb/in^2	3.5
Percent elongation	500
Brittle point °F	−75
Specific gravity	0.86
Maximum service temperature, °F	300
Shore A hardness	75
Dielectric constant	2.5
Dielectric strength, V/mil	800
Percent water absorption	0.5
Resistance to chemicals at 75°F:†	
Nonoxidizing acids (20% H_2SO_4)	S
Oxidizing acids (10% HNO_3)	Q
Aqueous salt solutions (NaCl)	S
Aqueous alkaline solutions (NaOH)	S
Polar solvents (C_2H_5OH)	S
Nonpolar solvents (C_6H_6)	Q
Water	S

*lb/(in^2 · 0.145) = KPa (kilopascals); ft · lb/(in of notch · 0.0187) = cm · N/cm of notch.
†S = satisfactory; Q = questionable; U = unsatisfactory.

permit cross-linking with sulfur, but not enough to cause it to become badly embrittled in the presence of ozone. IIR has excellent resistance to permeability by gases, and this resistance is enhanced by chlorination or bromination of IIR. Considerable quantities of these elastomers are used as inner linings for tires, but they are also used as caulking compounds, cable coatings, weather stripping, and coated

TABLE 6.4 Properties of Butyl Rubber

Property, units*	
Tensile strength, 10^3 lb/in^2	2.6
Percent elongation	700
Modulus, 10^3 lb/in^2	0.725
Specific gravity	0.92
Glass transition temperature, °F	−94
Resistance to chemicals at 75°F:†	
Nonoxidizing acids (20% H_2SO_4)	S
Oxidizing acids (10% HNO_3)	Q
Aqueous salt solutions (NaCl)	S
Aqueous alkaline solutions (NaOH)	S
Polar solvents (C_2H_5OH)	S
Nonpolar solvents (C_6H_6)	U
Water	S

*lb/(in^2 · 0.145) = KPa (kilopascals).
†S = satisfactory; Q = questionable; U = unsatisfactory.

fabrics. The loss in air pressure of inner tubes after 1 month was 50 percent for NR and 7 percent for IIR. The loss in pressure with chlorobutyl and bromobutyl rubber was almost undetectable under these same test conditions.

The properties of IIR are shown in Table 6.4.

6.10 Fluorinated Elastomers

The fluorocarbon and fluorosilicone elastomers are expensive polymers which have excellent resistance to hot organic liquids. PTFE is one of the best materials of construction for high-temperature corrosive environments. Difficulties associated with the fabrication of PTFE have been overcome by replacing one or more of the fluorine pendant groups by hydrogen or chlorine atoms or by copolymerization with monomers, such as ethylene.

While these modified polymers can be molded and extruded, they cannot be used in applications which require an elastic material. However, a series of elastomeric polyfluorocarbons has been produced by M. W. Kellogg, Du Pont, Asahi, and 3M.

The first commercial fluorinated elastomers were the copolymers of vinylidene fluoride ($H_2C{=}CF_2$) and chlorotrifluoroethylene ($F_2C{=}CFCl$; Kel-F) and poly-l,l-dihydroperfluorobutyl acrylate (IF4). However, IF4 had limited use because of its lack of resistance to corrosives and steam.

A copolymer of vinylidene fluoride and hexafluoropropylene [$F_2C{=}CF(CF_3)$] was introduced by 3M and Du Pont in the 1950s under the trade names Fluorel and Viton. These continue to be the most widely used fluoroelastomers. A terpolymer called Viton B was developed later by Du Pont by the incorporation of TFE with the monomer used to produce Viton. Copolymers of perfluoromethylvinyl ether [$F_2C{=}CF(OCF_3)$] and vinylidene fluoride or TFE were introduced by Du Pont in the 1970s under the trade name Kalrez.

More recently, Asahi has introduced terpolymers of TFE, propylene, and an other undisclosed monomer, under the trade name Aflas (TFE/P and TFE/PH). The properties of copolymers of vinylidene fluoride and hexafluoropropylene are shown in Table 6.5.

A typical fluorosilicone elastomer has a methyl and a trifluoropropyl ($-CH_2CH_2CF_3$) pendant group on each silicon atom in the siloxane repeating units of the polymer chain. It is customary to reinforce these polymers with fumed or precipitated silica and to cure them with organic peroxides. The properties of a fluorosilicone elastomer (Silastic) are shown in Table 6.6.

TABLE 6.5 Properties of a Typical Cured Copolymer of Vinylidene Fluoride and Hexafluoropropylene

Property, units*	
Tensile strength, 10^3 lb/in^2	
At 75°F	2.4
At 500°F	0.3
Percent elongation	
At 75°F	300
At 500°F	80
Tensile modulus, 10^3 lb/in^2	400
Brittle point °F	−30
Shore A hardness	
At 75°F	75
At 500°F	63

*lb/(in^2 · 0.145) = KPa (kilopascals).

TABLE 6.6 Properties of a Typical Fluorosilicone Elastomer

Property, units*	
Tensile strength, 10^3 lb/in^2	0.6
Percent elongation	600
Brittle point °F	−90
Shore A hardness	68
Resistance to chemicals at 75°F:†	
Nonoxidizing acids (20% H_2SO_4)	S
Oxidizing acids (10% HNO_3)	S
Aqueous salt solutions (NaCl)	S
Aqueous alkaline solutions (NaOH)	S
Polar solvents (C_2H_5OH)	S
Nonpolar solvents (C_6H_6)	S
Water	S

*lb/(in^2 · 0.145) = KPa (kilopascals).
†S = satisfactory; Q = questionable; U = unsatisfactory.

6.11 Elastomeric Copolymers of Styrene and Butadiene (SBR)

Unreinforced methyl rubber, a polymer of 2,3-dimethylbutadiene, proved to be inferior to NR when used for tires on German military trucks during World War I. It is satisfactory for making tires when reinforced with carbon black, but prior to World War II German chemists R. E. Tschunkur, W. Bock, and E. Konrad transferred their interests to copolymers of butadiene, which are produced by free-radical polymerization in aqueous emulsion, rather than methyl rubber, which

TABLE 6.7 Properties of a Typical Copolymer of Styrene (30) and Butadiene (70)

Property, units*	
Tensile strength, 10^3 lb/in^2	1.6
Percent elongation	400
300% modulus, 10^3 lb/in^2	1.4
Shore A hardness	65
Specific gravity	1.25
Percent water absorption	1.2
Resistance to chemicals at 75°F:†	
Nonoxidizing acids (20% H_2SO_4)	S
Oxidizing acids (10% HNO_3)	U
Aqueous salt solutions (NaCl)	S
Aqueous alkaline solutions (NaOH)	S
Polar solvents (C_2H_5OH)	S
Nonpolar solvents (C_6H_6)	U
Water	S

*lb/(in^2 · 0.145) = KPa (kilopascals).
†S = satisfactory; Q = questionable; U = unsatisfactory.

was produced by sodium ion anionic polymerization. Polydimethylbutadiene, which can also be produced in emulsion-polymerization systems, can be used as an additive to improve the strength of unreinforced polymers of isoprene.

Elastomers based on copolymers of styrene (30) and 1,3-butadiene (70), called Buna S, were patented in Germany in the 1930s and produced in large quantities in the United States under the acronym GRS during World War II. GRS (hot rubber), which was produced by emulsion polymerization at 49°C (120°F), was inferior to the comparable copolymer produced at lower temperature (cold rubber) after the armistice. The "cold rubber," which is produced by free-radical polymerization of styrene and butadiene in an aqueous emulsion at 5°C (41°F), is now called SBR. SBR is a major elastomer used throughout the world; it is somewhat inferior to NR but is used in place of NR in tires, belts, and mechanical goods.

The properties of SBR are shown in Table 6.7.

6.12 Elastomeric Copolymers of Acrylonitrile and Butadiene (NBR)

Random copolymers of acrylonitrile and butadiene (NBR) are produced by free-radical polymerization in aqueous emulsions using techniques similar to those used for the production of SBR. NBR is more resistant to organic solvents than SBR, and this improved resistance is related directly to the AN content in NBR. Commercial elastomers

TABLE 6.8 Properties of a Typical Acrylonitrile-Butadiene Elastomer (NBR)

Property, units*	
Tensile strength, 10^3 lb/in^2	2.7
Percent elongation	300
Shore A hardness	72
Resistance to chemicals at 75°F:†	
Nonoxidizing acids (20% H_2SO_4)	S
Oxidizing acids (10% HNO_3)	U
Aqueous salt solutions (NaCl)	S
Aqueous alkaline solutions (NaOH)	S
Polar solvents (C_2H_5OH)	S
Nonpolar solvents (C_6H_6)	Q
Water	S

*lb/(in^2 · 0.145) = KPa (kilopascals).
†S = satisfactory; Q = questionable; U = unsatisfactory.

(Hycar and Chemigum) with as little as 20 percent AN (low) and as high as 40 percent AN (high) are available commercially.

Vulcanized NBR is used for grease seals, O-rings, gaskets, and hose at temperatures as high as 121°C (250°F) in air and as high as 149°C (300°F) when immersed in oil.

The properties of a typical NBR are shown in Table 6.8.

6.13 TPEs

Unvulcanized NR is a thermoplastic elastomer, but because of excessive flow, it is not particularly useful unless the flow is controlled physically, as in the Mackintosh cloth laminate, or chemically by the introduction of sulfur cross-links. In the 1950s, polymer scientists recognized that elasticity is independent of the chemical structure of the polymer molecule, but dependent on the ability of a polymer chain to undergo reversible stretching and contraction. This information made possible the tailoring of elastomers in which the excessive flow is restricted physically rather than by chemical cross-links, i.e., the making of TPEs.

The requirement for a TPE is that it be a multiphase polymer at ordinary temperatures and that one phase be rubbery and the other phase be hard. It is also required that the hard phase soften when the temperature is increased.

The most widely used TPE is the triblock of polystyrene-polybutadiene-polystyrene in which the polybutadiene is the soft elastic domain and the polystyrene is the hard, thermoplastic domain. The worldwide and U.S. production of this SBS block terpolymer (Kraton) in 1987 was 250,000 and 125,000 t, respectively. This TPE and an-

TABLE 6.9 Properties of a Typical Styrene-Butadiene-Styrene Thermoplastic Elastomer

Property	Kraton 30% SM	Kraton 14% SM
Tensile strength, 10^3 lb/in^2*	4.6	3.1
Percent elongation	880	1300
300% modulus, 10^3 lb/in^2*	0.4	0.1
Shore A hardness	71	37
Specific gravity	0.94	0.92
Resistance to chemicals at 75°F:†		
Nonoxidizing acids (20% H_2SO_4)	S	S
Oxidizing acids (10% HNO_3)	U	U
Aqueous salt solutions (NaCl)	S	S
Aqueous alkaline solutions (NaOH)	S	S
Polar solvents (C_2H_5OH)	S	S
Nonpolar solvents (C_6H_6)	U	U
Water	S	S

*lb/(in^2 · 0.145) = KPa (kilopascals).
†S = satisfactory; Q = questionable; U = unsatisfactory.

other SB block copolymer (K-Resin) are produced by anionic-polymerization techniques in which a controlled quantity of the second monomer is added to the stable macroanion produced by the polymerization of the first polymer.

Isoprene may be used in place of butadiene in Kraton, and this TPE may be hydrogenated to produce a block copolymer with improved weatherability. Kraton may be readily injection-molded, and the molded parts, such as shoe soles, may be used in place of vulcanized NR or SBR. The properties of SBS thermoplastic elastomers are shown in Table 6.9.

The world and U.S. production of blends of ethylene-propylene random copolymer (EPM) and polypropylene (it PP) in 1987 was 150,000 and 75,000 t, respectively. These TPEs, which are sold under the trade names of Santoprene, Somel, and Telcar, are also used in place of vulcanized NR or vulcanized SBR for wire insulation and automotive parts.

The stress-strain curves for Santoprene (Shore A hardness 64 and 73) at various temperatures are shown in Figs. 6.3 and 6.4. The dynamic mechanical properties of Santoprene as a function of temperature are shown in Figs. 6.5 and 6.6. These data were supplied by Monsanto. Other properties of these TPEs are shown in Table 6.10.

In 1987, 60,000 and 30,000 t of thermoplastic urethane elastomer (TPU) were produced worldwide and in the United States, respectively. These abrasion-resistant TPEs consist of flexible polyester blocks and hard crystalline polyurethane blocks. These multiblock TPUs are used in automotive and biomedical applications as well as

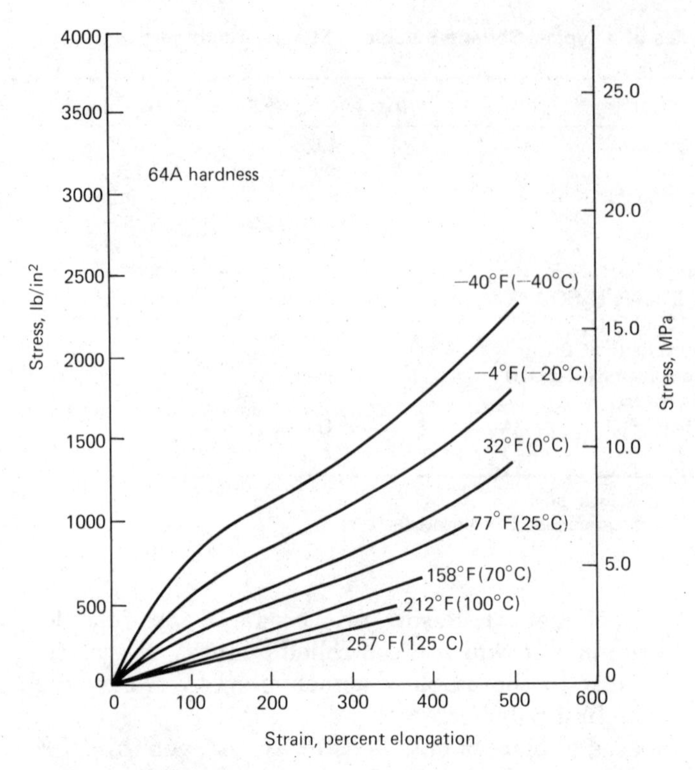

Figure 6.3 Stress-strain curves for Santoprene at various temperatures at 64A hardness.

for mechanical goods and fabric coatings. These TPUs are sold under the trade names Estane, Texin, and Pellathane.

The properties of a typical TPU are shown in Table 6.11.

Thermoplastic polyesters (Hytrel) are produced by the tetrabutyl titanate–catalyzed transesterification of a mixture of dimethyl terephthalate, polyether glycol, and an excess of 1,4-butanediol. These TPEs are used in place of vulcanized rubber for belts, shoe soles, wire coatings, hose, and automotive parts. The properties of these thermoplastic polyesters are shown in Table 6.12.

Polyether block amide TPEs (Pebax) were introduced commercially in 1981. These TPEs consist of a hard domain of a polyamide (PA) and a soft domain of a polyether. The properties of Pebax can be varied by choice of the polyamide used. These injection-moldable TPEs with tensile strengths ranging from 4000 to 8000 lb/in^2 are available. The TPEs are being used for athletic footwear, knee pads, and molded flexible articles.

The properties of typical polyether block amides are shown in Table 6.13.

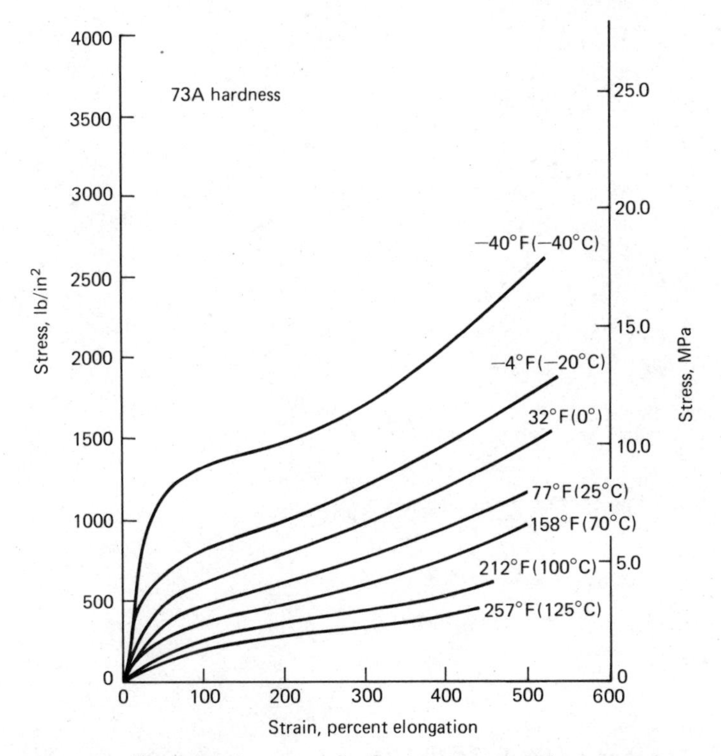

Figure 6.4 Stress-strain curves for Santoprene at various temperatures at 73A hardness.

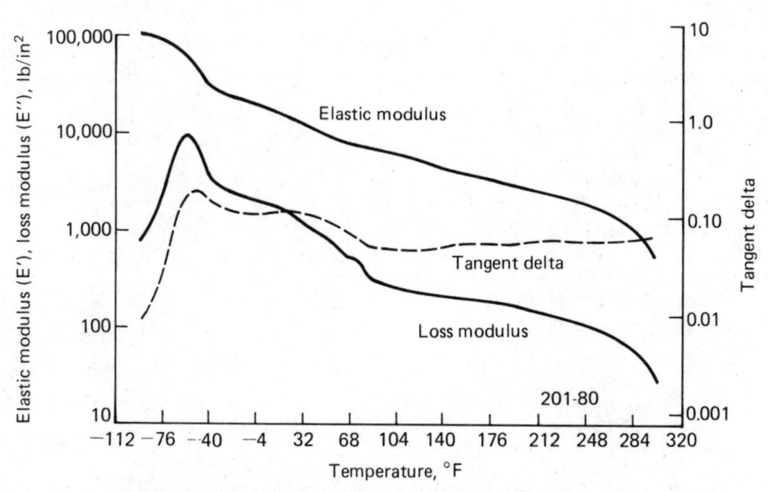

Figure 6.5 Mechanical properties of Santoprene as a function of temperature.

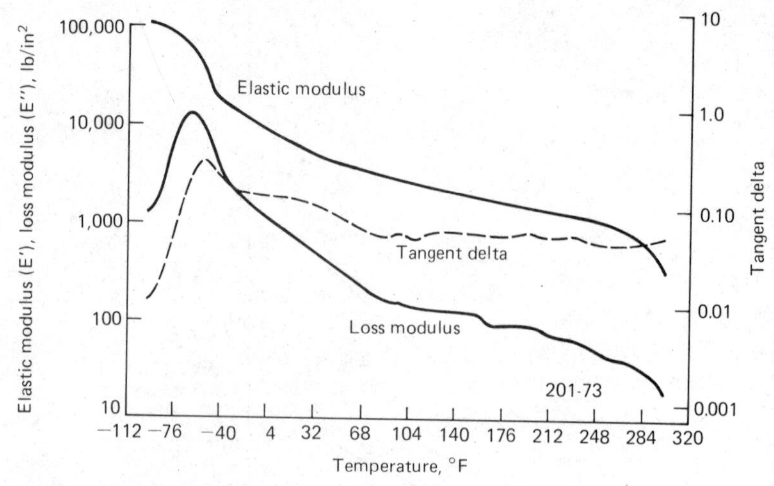

Figure 6.6 Mecahnical properties of Santoprene.

TABLE 6.10 Properties of a Typical TPE Blend of EPM and PP

Property, units*	
Tensile strength, 10^3 lb/in^2	2
Percent elongation	300
100% modulus, 10^3 lb/in^2	1.5
Shore A hardness	70
Specific gravity	0.9
Resistance to chemicals at 75°F:†	
Nonoxidizing acids (20% H_2SO_4)	S
Oxidizing acids (10% HNO_3)	Q
Aqueous salt solutions (NaCl)	S
Aqueous alkaline solutions (NaOH)	S
Polar solvents (C_2H_5OH)	S
Nonpolar solvents (C_6H_6)	Q
Water	S

*lb/(in^2 · 0.145) = KPa (kilopascals).
†S = satisfactory; Q = questionable; U = unsatisfactory.

TABLE 6.11 Properties of a Typical Thermoplastic Polyurethane

Property, units*	
Tensile strength, 10^3 lb/in^2	4
Percent elongation	500
100% modulus, 10^3 lb/in^2	1.5
Shore A hardness	85
Specific gravity	1.2
Resistance to chemicals at 75°F:†	
Nonoxidizing acids (20% H_2SO_4)	Q
Oxidizing acids (10% HNO_3)	U
Aqueous salt solutions (NaCl)	S
Aqueous alkaline solutions (NaOH)	Q
Polar solvents (C_2H_5OH)	U
Nonpolar solvents (C_6H_6)	Q
Water	S

*lb/(in^2 · 0.145) = KPa (kilopascals).
†S = satisfactory; Q = questionable; U = unsatisfactory.

TABLE 6.12 Properties of a Typical Thermoplastic Polyester

	Percent hard segment		
Property, units*	33	58	76
Tensile strength, 10^3 lb/in^2	5.7	6.4	6.9
Percent elongation	810	760	510
Flexural modulus, 10^3 lb/in^2	6.5	30	72
Shore D hardness	55	60	72
Specific gravity	1.15	1.20	1.22
Resistance to chemicals at 75°F:†			
Nonoxidizing acids (20% H_2SO_4)	S	S	S
Oxidizing acids (10% HNO_3)	U	U	U
Aqueous salt solutions (NaCl)	S	S	S
Aqueous alkaline solutions (NaOH)	Q	Q	Q
Polar solvents (C_2H_5OH)	S	S	S
Nonpolar solvents (C_6H_6)	U	U	U
Water	S	S	S

*lb/(in^2 · 0.145) = KPa (kilopascals).
†S = satisfactory; Q = questionable; U = unsatisfactory.

TABLE 6.13 Properties of Typical Polyether Block Amides (Pebax)

Property, units*	Extrusion grade	Molding grade
Tensile strength, 10^3 lb/in^2	5	7.4
Percent elongation	700	510
Flexural modulus, 10^3 lb/in^2	3	30
Shore D hardness	35	55
Specific gravity	1.01	1.06
Percent water absorption	1.2	3.5
Resistance to chemicals at 75°F:†		
Nonoxidizing acids (20% H_2SO_4)	S	S
Oxidizing acids (10% HNO_3)	U	U
Aqueous salt solutions (NaCl)	S	S
Aqueous alkaline solutions (NaOH)	U	U
Polar solvents (C_2H_5OH)	S	S
Nonpolar solvents (C_6H_6)	S	S
Water	S	S

*lb/(in^2 · 0.145) = KPa (kilopascals).
†S = satisfactory; Q = questionable; U = unsatisfactory.

6.14 Glossary

accelerator A catalyst used to speed up the sulfur vulcanization of rubber.

acid, oxidizing An acid like nitric or chromic acid which is capable of oxidizing organic compounds.

ACM An acrylic elastomer.

acrylic elastomer One containing repeating units of an acrylic ester ($H_2C{=}CHCOOR$).

acrylonitrile $H_2C{=}CH(CN)$.

Aflas A terpolymer of tetrafluoroethylene, propylene, and another monomer.

amorphous Noncrystalline.

aqueous A water system.

blend A mixture.

BR Polybutadiene.

brittle point The temperature at which a polymer becomes brittle and is easily fractured as the temperature is decreased.

Buna A name used for copolymers of butadiene and styrene (Buna S) or acrylonitrile (Buna N).

Butadiene $H_2C{=}CH{-}CH{=}CH_2$.

1,4-Butanediol $HO(CH_2)_4OH$.

butyl rubber A copolymer of isobutylene with a small amount (5%) of isoprene.

carbon black Finely divided amorphous carbon particles produced by the incomplete combustion of a hydrocarbon.

caulking material A sealant.

Chemigum NBR.

2-Chlorobutadiene Chloroprene.

2-Chloroethyl acrylate $H_2C{=}CHCOOC_2H_4Cl$.

Chloroprene 2-chlorobutadiene, $H_2C{=}CCl{-}CH{=}CH_2$.

cis Having the pendant groups or chain extensions on the same side of the double bond ($-C{=}C-$) in an unsaturated isomer.

cold flow Dimensional change over a period of time, i.e., creep.

conformation Shapes of molecular chains.

copolymer A polymer consisting of two or more repeating units.

copolymer, block A copolymer consisting of long sequences of the same repeating units.

copolymer, random A copolymer in which the repeating units are not in a specific order.

CR Neoprene.

cross-link density A qualitative measure of the extent of cross-linking.

cross-links Covalent bonds between polymer chains.

D Diene; two ethylenic double bonds ($C{=}C$).

ΔS Change in entropy.

elastomer Rubber.

elastomer, thermoplastic A thermoplastic consisting of soft and hard segments (domains).

emulsion A dispersion of a substance in water.

entropy A measure of the extent of disorder, randomness.

EPDM Vulcanizable ethylene-propylene rubber; vulcanizable EPM.

EPM Ethylene-propylene copolymer.

epoxidized rubber NR in which the ethylenic carbon atoms are joined by an epoxy group.

Estane A TPU.

ester RCOOR′.

extrusion A process in which a heat-softened thermoplastic is forced through an orifice continuously.

filler, reinforcing A filler which improves the strength of a polymer.

Florel A perfluorocarbon elastomer.

fluorosilicone A silicone with a trifluoropropyl pendant group on each repeating siloxane unit.

free radical An electron-deficient atom or molecule (R ·).

glass transition temperature T_g The temperature at which a glasslike solid becomes somewhat flexible as the temperature is increased.

Goodyear, Charles The inventor of vulcanized rubber.

Gough-Joule effect The increase in temperature during the stretching of an elastomer.

GRS Buna S, SBR.

gum stock An unfilled elastomer,

Heveaplus A copolymer of NR and methyl methacrylate.

hexafluoropropylene $F_2C{=}CF(CF_3)$.

hookean Obeys Hooke's law.

Hooke's law Stress is proportional to strain.

Hycar An NBR.

Hytrel A TPE, polyester.

IF4 A poly(1,1-dihydrotrifluoroethylene).

intermolecular forces Attractive forces between polymer chains.

IIR Butyl rubber.

intermolecular forces Attractive forces between polymer chains.

IR Polyisoprene.

isobutylene $H_2C{=}C(CH_3)_2$.

isoprene 2-methylbutadiene: $H_2C{=}C(CH_3){-}CH{=}CH_2$.

it PP Isotactic PP, one with all methyl pendant groups on the same side of the polymer chain.

Kalrez A copolymer of perfluorovinyl ether and vinylidene fluoride.

Kel-F A trade name for a polymer or copolymer of trifluorochloroethylene.

ketone $R_2C{=}O$.

Kraton A TPE, a block copolymer of styrene-butadiene-styrene.

K-Resin A block copolymer of styrene and butadiene.

laminate A sheet consisting of several layers.

London forces Weak, intermolecular (van der Waals) forces.

Mackintosh A laminate of NR and a fabric, named after its inventor, Charles Mackintosh.

macromolecule A polymer, a giant molecule.

methyl group —CH_3.

methyl rubber Polydimethylbutadiene.

Modulus 300% The modulus at 300 percent elongation.

modulus Stress/strain.

Modulus, Young's The ratio of tensile stress to tensile strain.

molding, injection A process in which a heat-softened thermoplastic is converted to a solid by forcing it through a cylinder into a cooled cavity and ejecting the finished cooled article.

NBR Acrylonitrile-butadiene rubber.

neoprene A polymer of chloroprene, 2-chlorobutadiene.

norbornene A cyclic cross-linkable monomer.

NR Natural rubber.

ozone O_3.

PA Polyamide.

PB Polybutadiene rubber.

Pebax A polyether block amide.

Pellathane A TPU.

pendant group A functional group on a polymer chain, such as the methyl group on polypropylene: $-\!(CH_2—C(CH_3)H)\!-$.

perfluoro Multiple fluorine groups.

permeability The ability of gases or liquids to pass through a membrane.

plasticizer A softener.

Poisson ratio, ν The ratio of the transverse contracting strain to the elongation strain of a stretched rod.

polyamide $-\!(HNRNHCORCO)\!-$.

polyester $-\!(ORCOOR)\!-$.

polyether $-\!(ROR)\!-$.

polyfluorocarbon A polymer containing fluorine (F) atoms.

PTFE Polytetrafluoroethylene (Teflon).

rubber, cold SBR produced at relatively low temperatures [5°C (41°F)].

rubber, hard Highly cross-linked rubber (ebonite).

rubber, hot SBR produced at relatively high temperatures [45°C (120°F)].

S Entropy

Santoprene A TPE, a blend of EPM and it PP.

SBR Styrene-butadiene rubber.

SBS Kraton.

Sealant A void filler usually adhered to the void surface.

section, principal Segments of polymer chains between cross-links.

Shore A hardness The indentation hardness measured by a blunt needle or by use of a scleroscope, which measures the rebound when a conical hammer strikes the test specimen.

SI Silicone.

Silastic A silicone elastomer.

silica SiO_2.

silica, fumed Finely divided silica obtained by the reaction of silicon tetrachloride ($SiCl_4$) and hydrogen (H_2).

silicone $\leftarrow SiR_2O \rightarrow$.

simulated structure A simplified skeletal structural representation.

sodium ion Na^+.

solvent, nonpolar A solvent such as aliphatic or aromatic hydrocarbons.

solvent, polar A solvent such as water, alcohol, or an amine.

Somel A TPE, a blend of EPM and it PP.

SPR Single-ply roofing.

SR Synthetic rubber.

strain Elongation.

stress A force acting on an object, such as tensile or stretching force.

styrene $H_2C{=}CH(C_6H_5)$.

Telcar A TPE, a blend of EPM and it PP.

terpolymer A copolymer consisting of three repeating units.

Texin A TPU.

T_g Glass transition temperature.

Thiokol A polyethylene sulfide, solvent-resistant elastomer, the first American synthetic rubber.

TPE Thermoplastic elastomer.

TPU Thermoplastic polyurethane.

trans Having the pendant groups or chain extension on the opposite side of the double bond $\leftarrow C{=}C \rightarrow$ in an unsaturated isomer.

transesterification The replacement of one alcohol group in an ester by another alcohol group.

transverse contraction Contraction perpendicular to the axis of a rod.

Tschunkur, R. The coinventor of Buna S and Buna N.

unsaturated Containing ethylenic (C=C) groups.

urethane —RNHCOOR′—.

vinyl group —CH=CH_2.

vinylidene fluoride H_2C=CF_2.

Viton A perfluorocarbon.

vulcanized Cross-linked.

zinc oxide ZnO.

6.15 References

Allen, P. W., P. B. Lindley, and A. R. Payne: *Use of Rubber in Engineering,* Maclaren & Sons Ltd., London, 1967.

Alliger, G., and I. J. Sjothun: *Vulcanization of Elastomers,* Reinhold Publishing Corporation, New York, 1964.

Bertran, H. H.: *Developments in Rubber Technology: Synthetic Rubbers,* Applied Science Publishers Ltd., London, 1981.

Holden, G.: "Elastomers Thermoplastic," in J. I. Kroschwitz (ed.): *Encyclopedia of Polymer Science and Engineering,* vol. 5, John Wiley & Sons, Inc., Interscience Publishers, New York, 1986.

Hull, D. E.: *Elastomerics* 119(11):12 (1987).

Kennedy, J. P., and E. G. M. Tornqvist (eds.): *Polymer Chemistry of Synthetic Elastomers,* John Wiley & Sons, Inc., New York, 1968.

Meals, R. N., and F. M. Lewis: *Silicones,* Reinhold Publishing Corporation, New York, 1959.

Morton, M. (ed.): *Rubber Technology,* Van Nostrand Reinhold Company, Inc., New York, 1973.

Ogintz, S.: *Elastomerics* 119(11):21 (1987).

Rochow, E. G., *Introduction to the Chemistry of Silicones,* John Wiley & Sons, Inc., New York, 1951.

Saltman, W. M. (ed.): *Stereo Rubbers,* John Wiley & Sons, Inc., New York, 1977.

Seymour, R. B., and G. S. Kirshenbaum (eds.): *High Performance Polymers: Their Origin and Development,* Elsevier Science Publishing Company, Inc., New York, 1986.

Seymour, R. B., and C. E. Carraher: *Polymer Chemistry: An Introduction,* Marcel Dekker, Inc., New York, 1988.

Tate, D. P., and T. W. Bethea: "Butadiene Polymers," in J. I. Kroschwitz (ed.): *Encyclopedia of Polymer Science and Engineering,* vol. 2, John Wiley & Sons, Inc., Interscience Publishers, New York, 1985.

Tornqvist, E. G. M.: in R. B. Seymour, and T. Cheng (eds.): *History of Polyolefins,* Reidel Publishing Co., Dordrecht, The Netherlands, 1986.

Whitby, G. S., C. C. Davis, and R. F. Dunbrook (eds.): *Synthetic Rubber,* John Wiley & Sons, Inc., New York, 1954.

Chapter

7

Styrene Polymers

7.1 Introduction

PS was produced by Neuman in the last part of the nineteenth century and was commercialized over a century later by I. G. Farbenindustrie and Dow in the early 1930s. In 1987, 2.18 million t of this polymer was produced in the United States. Over 1.27 million t of PS was used in building and construction, but the softening point of this clear brittle polymer is too low to meet the specifications of an engineering polymer. The impact resistance of PS is improved by blending with elastomers, but these toughened plastics cannot be used at high temperatures, i.e., above 88°C (190°F).

Over 45,000 t of a copolymer of styrene ($CH_2{=}CHCC_6H_5$) and AN ($H_2C{=}CHCN$) was produced in the United States in 1987. This copolymer (SAN) has a higher softening point [99°C (210°F)] and is more resistant to aliphatic solvents than PS, but it is not classified as an engineering plastic.

In 1941, Seymour and Kispersky produced an engineering polymer by the copolymerization of styrene, AN, and maleic anhydride. This terpolymer (Cadon) has been toughened by the addition of elastomers, and its utility has been increased by blending with other polymers.

In 1948, Daley toughened SAN by blending it with NBR. In 1953, Calvert blended emulsions of these copolymers and in 1957 patented a graft copolymer of AN, butadiene, and styrene (ABS). In 1987, 545,000 t of ABS was produced in the United States, and 62,000 t of this terpolymer was used for pipe and fittings.

7.2 SAN

SAN copolymers are characterized by excellent gloss and good thermal and chemical resistance. The dimensional stability and heat resistance of SAN is improved by reinforcing with fiberglass.

Olefin-modified SAN (OSA) is produced by the polymerization of

TABLE 7.1 Properties of Typical Styrene-Acrylonitrile Copolymer

Property, units*	SAN	OSA	20% glass-filled SAN
Melting point T_m, °F	—	—	—
Glass transition temp. T_g, °F	120	75	120
Processing temp., °F	350	425	475
Molding pressure, 10^3 lb/in^2	10	1	15
Mold shrinkage, 10^{-3} in/in	4	6	2
Heat deflection temp. under flexural load of 264 lb/in^2, °F	215	200	220
Maximum resistance to continuous heat, °F	180	160	190
Coefficient of linear expansion, 10^{-6} in/(in · °F)	35	40	15
Compressive strength, 10^3 lb/in^2	14	10	20
Izod impact strength, ft · lb/in of notch	0.5	14	1.0
Tensile strength, 10^3 lb/in^2	11	6	25
Flexural strength, 10^3 lb/in^2	12	8.5	21
Percent elongation	2	20	1.5
Tensile modulus, 10^3 lb/in^2	500	300	1500
Flexural modulus, 10^3 lb/in^2	550	300	1500
Rockwell hardness	R80	R100	R120
Specific gravity	1.07	1.02	1.30
Percent water absorption	0.5	0.4	0.7
Dielectric constant	3	3	3
Dielectric strength, V/mil	425	425	500
Resistance to chemicals at 75°F:†			
Nonoxidizing acids (20% H_2SO_4)	S	S	S
Oxidizing acids (10% HNO_3)	Q	Q	Q
Aqueous salt solutions (NaCl)	S	S	S
Polar solvents (C_2H_5OH)	S	S	S
Nonpolar solvents (C_6H_6)	U	U	U
Water	S	S	S
Aqueous alkaline solutions (NaOH)	S	S	S

*lb/(in^2 · 0.145) = KPa (kilopascals); ft · lb/(in of notch · 0.0187) = cm · N/cm of notch.
†S = satisfactory; Q = questionable; U = unsatisfactory.

styrene and acrylonitrile in the presence of an olefin elastomer such as EPM. OSA, which has a higher impact resistance than SAN, may be blended with ABS, PVC, or PC. These and other AN polymers discussed in this chapter can be injection-molded and extruded to produce molded articles or profiles, including pipe. The properties of typical SANs are shown in Table 7.1.

SAN is available both as an unfilled copolymer and with 20, 30, and 45 percent fiberglass. Some reinforced SAN composites are lubricated by silicones (2%). Composites with chopped and long glass fibers are available. Dow and Monsanto also market weather-resistant SAN. The principal sources and trade names for SAN are shown here:

Principal source	Trade name
BASF	Luran
Comalloy	240-3020
Complas	RSA, 2020
Dow	Rovel, Tyril
Federal	FPC 818
LNP	B-1000
Monsanto	Lustron
RTP	RTP 501
A. Schulman	Polyman 551
Thermofil	B-20FG-0100
Wilson-Fiberfil	G 40/20

7.3 Styrene-Maleic Anhydride Copolymers

Terpolymers of styrene, maleic anhydride, and AN (SMA-AN; Cadon) and numerous blends of these terpolymers and other polymers are produced by Monsanto. In some cases, the AN is replaced by acrylic esters and isobutylene. Comparable polymer products (Dylark) are also produced by ARCO.

The good impact resistance of these polymeric products can be increased by increasing the AN content in the terpolymer. The T_g as well as the heat deflection temperature can be increased by increasing the maleic anhydride content of the terpolymer. The properties of typical SMA-AN polymers are shown in Table 7.2.

7.4 SMA-PC Alloys

Alloys with and without fiberglass reinforcement have been produced by blending SMA copolymers or terpolymers with PC. Glass-reinforced alloys with 10 and 20 percent fiberglass are available from ARCO under the trade name Arloy. BASF produces a high-gloss alloy under the trade name Styrolux; Dow produces a weather-resistant product, and Firestone produces an FDA-approved product under the trade name Stereon.

7.5 High-Barrier Polymers

Polyacrylonitrile (PAN) has outstanding resistance to permeability by gases and moisture vapor, but it is difficult to mold or extrude this brittle polymer. However, solvent cast-oriented films can be used as barrier films.

The intractability of PAN has been overcome by the production of

TABLE 7.2 Properties of Typical Styrene-Maleic Anhydride–Based Polymers

Property, units*	SMA-AN	Elastomer-modified SMA-AN	20% fiberglass-reinforced SMA-AN
Melting point T_m, °F	—	—	—
Glass transition temp. T_g, °F	—	—	—
Processing temp., °F	400	450	500
Molding pressure, 10^3 lb/in^2	15	15	20
Mold shrinkage, 10^{-3} in/in	5	5	2.5
Heat deflection temp. under flexural load of 264 lb/in^2, °F	225	210	235
Maximum resistance to continuous heat, °F	200	190	210
Coefficient of linear expansion, 10^{-6} in/(in · °F)	45	60	30
Compressive strength, 10^3 lb/in^2	—	—	—
Izod impact strength, ft · lb/in of notch	1.2	3.5	2.0
Tensile strength, 10^3 lb/in^2	8	5	15
Flexural strength, 10^3 lb/in^2	12	8	14
Percent elongation	6	20	2.5
Tensile modulus, 10^3 lb/in^2	400	300	800
Flexural modulus, 10^3 lb/in^2	400	300	800
Rockwell hardness	R107	R85	R75
Specific gravity	1.06	1.06	1.14
Percent water absorption	0.1	0.3	0.1
Dielectric constant	3	3	3
Dielectric strength, V/mil	400	450	450
Resistance to chemicals at 75°F:†			
Nonoxidizing acids (20% H_2SO_4)	S	S	S
Oxidizing acids (10% HNO_3)	Q	Q	Q
Aqueous salt solutions (NaCl)	S	S	S
Polar solvents (C_2H_5OH)	S	S	S
Nonpolar solvents (C_6H_6)	U	U	U
Water	S	S	S
Aqueous alkaline solutions (NaOH)	S	S	S

*lb/(in^2 · 0.145) = KPa (kilopascals); ft · lb/(in of notch · 0.0187) = cm · N/cm of notch.
†S = satisfactory; Q = questionable; U = unsatisfactory.

useful copolymers with a high AN content. The most widely used AN-type barrier films are copolymers of AN—methyl acrylate and NBR (Barex) produced by Vistron, and a styrene-AN copolymer composition (Lopac) produced by Monsanto. Both these resins have been used to mold bottles for carbonated beverages, but the Food and Drug Administration (FDA) questioned this end use because of the possibility of absorption of AN into the contents of the containers. Tests on laboratory animals have shown that AN is a carcinogen.

Nevertheless, the FDA approved the use of Barex film in contact with foods, such as wheat, cheese, and snack foods, as well as pharma-

TABLE 7.3 Properties of AN Polymeric Films

Property, units	PAN	Lopac	Barex 210
Heat deflection temperature under flexural load of 264 lb/in^2, °F	—	203	158
Maximum resistance to continuous heat, °F	—	175	145
Tensile strength, 10^3 lb/in^2	—	6	3
Percent elongation	—	2	150
Flexural modulus, 10^3 lb/in^2	—	11	49
Specific gravity	1.17	1.15	1.15
Permeability to CO_2 at 445°F, 10^{-17} mol/(Pa · s · m)	0.75	3.75	4.0
Permeability to moisture at 445°F, 10^{-17} mol/(Pa · s · m)	—	95	70

ceuticals. Some data on these polymeric products are outlined in Table 7.3.

7.6 ABS Terpolymer

ABS terpolymers are available as a variety of blends and graft copolymers. Most ABS products consist of grafted polybutadiene particles as the discontinuous phase dispersed in a continuous phase of SAN. The SAN composition is usually 70 percent styrene and 30 percent AN. ABS contains 70 to 90 percent SAN.

The impact resistance of ABS is a function of the butadiene content, and the tensile strength and heat resistance are functions of the SAN content. ABS is notch-sensitive but not as sensitive as nylon or PC. It retains its good impact resistance over a wide temperature range. This amorphous polymer is also characterized by good resistance to creep.

ABS may be injection-molded and extruded, and sheets of ABS may be readily thermoformed. When the thermoformed article is subject to stress, it is advantageous to include fillets with a radius equal to the wall thickness.

The major markets for ABS are appliances (18.5%), building and construction (17.5%), transportation (16%), business machines, etc. (15%), recreation (5.5%), packaging (2%), and luggage (2%). As shown in Fig. 7.1, large complex automotive parts, such as bumper fairings, are readily blow-molded from ABS.

Electroplated ABS is used as automotive grills, headlight bezels, wheel covers, and mirror housings. Thermoformed ABS is also used for refrigerator liners and recreational boats. Extruded ABS is used for both waste, vent, and drain pipe and for higher-pressure pipe.

ABS is available from Borg-Warner (GE), Monsanto, Dow, Bolcol,

Figure 7.1 Blow-molding of large complex parts. Cycolac brand ABS from Borg-Warner is used for blow-molding large complex parts like those illustrated here. These bumper fairings, used on the International® 8300 and 9700 series trucks from Navistar International, help make the vehicles especially aerodynamic. Cycolac ABS demonstrates versatility for blow-molding applications. Advantages include good ductility, modulus, and melt strength. Grade LXB, used in bumper fairings, is designed for painted exterior applications such as this and has a good heat deflection temperature. (*Photograph courtesy of Borg-Warner.*)

LNP, and RTP under the trade names Cycolac, Lustran, Magnum, Taitalac, EMI and RTP, respectively. Flame-resistant ABS and fiberglass-reinforced ABS are available from several of these suppliers. The properties of typical ABS products are shown in Table 7.4.

TABLE 7.4 Properties of Typical ABS Polymers

				Reinforced ABS	
Property, units*	ABS	High-impact ABS	20% FG	20% long FG	20% graphite
Melting point T_m, °F	—	—	—	—	—
Glass transition temp. T_g, °F	250	200	210	210	210
Processing temp., °F	400	350	400	450	450
Molding pressure, 10^3 lb/in^2	15	15	25	30	25
Mold shrinkage, 10^{-3} in/in	6	6	2	1	1
Heat deflection temp. under flexural load of 264 lb/in^2, °F	205	205	210	215	215
Maximum resistance to continuous heat, °F	190	190	195	200	200
Coefficient of linear expansion, 10^{-6} in/(in · °F)	35	50	15	15	10
Compressive strength, 10^3 lb/in^2	6	5	13	15	16
Izod impact strength, ft · lb/in of notch	3	7	1	2	1
Tensile strength, 10^3 lb/in^2	5	5	12	13	15
Flexural strength, 10^3 lb/in^2	9	8	17	20	23
Percent elongation	10	20	1.5	2	1.5
Tensile modulus, 10^3 lb/in^2	350	200	800	1000	1600
Flexural modulus, 10^3 lb/in^2	400	300	700	850	1550
Rockwell hardness	R110	R95	R105	M90	R110
Specific gravity	1.2	1.0	1.2	1.2	1.2
Percent water absorption	0.3	0.3	0.2	0.2	0.15
Dielectric constant	—	—	—	—	—
Dielectric strength, V/mil	425	425	450	450	—
Resistance to chemicals at 75°F:†					
Nonoxidizing acids (20% H_2SO_4)	S	S	S	S	S
Oxidizing acids (10% HNO_3)	Q	Q	Q	Q	Q
Aqueous salt solutions (NaCl)	S	S	S	S	S
Polar solvents (C_2H_5OH	Q	Q	Q	Q	Q
Nonpolar solvents (C_6H_6)	U	U	U	U	U
Water	S	S	S	S	S
Aqueous alkaline (NaOH)	S	S	S	S	S

*lb/(in^2 · 0.145) = KPa (kilopascals); ft · lb/(in of notch · 0.0187) = cm · N/cm of notch.
†S = satisfactory; Q = questionable; U = unsatisfactory.

7.7 Acrylic-Styrene-Acrylonitrile Terpolymers

Acrylic-styrene-acrylonitrile terpolymers (ASA), which are produced under the trade name Geloy by General Electric, have better weather resistance than ABS. The properties of this amorphous polymer are similar to those of ABS. ASA is also used to produce alloys with PVC, PC, and PMMA. The properties of a typical ASA terpolymer are shown in Table 7.5.

7.8 Other AN Copolymers

Showa-Denko is producing a weather-resistant terpolymer by the copolymerization of styrene and AN in the presence of chlorinated poly-

TABLE 7.5 Properties of Typical ASA Terpolymers

Property, units*	
Melting point T_m, °F	—
Glass transition temp. T_g, °F	—
Processing temp., °F	400
Molding pressure, 10^3 lb/in^2	12
Mold shrinkage, 10^{-3} in/in	3
Heat deflection temp. under flexural load of 264 lb/in^2, °F	185
Maximum resistance to continuous heat, °F	170
Coefficient of linear expansion, 10^{-6} in/(in · °F)	35
Compressive strength, 10^3 lb/in^2	6
Izod impact strength, ft · lb/in of notch	9
Tensile strength, 10^3 lb/in^2	4
Flexural strength, 10^3 lb/in^2	7
Percent elongation	30
Tensile modulus, 10^3 lb/in^2	300
Flexural modulus, 10^3 lb/in^2	230
Rockwell hardness	R85
Specific gravity	1.05
Percent water absorption	0.03
Dielectric constant	0.3
Dielectric strength, V/mil	400
Resistance to chemicals at 75°F:†	
Nonoxidizing acids (20% H_2SO_4)	S
Oxidizing acids (10% HNO_3)	Q
Aqueous salt solutions (NaCl)	S
Polar solvents (C_2H_5OH)	Q
Nonpolar solvents (C_6H_6)	U
Water	S
Aqueous alkaline solutions (NaOH)	S

*lb/(in^2 · 0.145) = KPa (kilopascals); ft · lb/(in of notch · 0.0187) = cm · N/cm of notch.

†S = satisfactory; Q = questionable; U = unsatisfactory.

ethylene. Because of the chlorine content, the molding and extrusion temperature must be below 210°C (410°F). ACS has better flame resistance than ABS or ASA and is used for molding office machine housings and electrical components.

7.9 Thermoforming

The first plastic materials, such as tortoise shells and animal horns, were formed into useful articles by draping the heated materials over appropriate forms. The ambiguous term "thermoforming" is still applied to the process of fabricating warmed plastic sheets, such as ABS. This term is used to describe the shaping of sheets with a thickness of less than 10 mils, i.e., films and thicker plastic sheets.

In the simplest process, cut-sheet is heated and formed by draping over a male form. The warmed sheet may be formed against the mold surface by pressing by hand, by mechanical means, or by compressing with air pressure. In all versions of the molding process, the warmed sheet must be allowed to cool before removal from the mold.

The trend is to reduce the temperature of the sheet and to employ higher pressure. The ultimate, in which the sheet is not heated at all, is called solid-phase forming. ABS and ABS/PC are readily formed at room temperature in a process called closed-die forming in which male and female molds are made by closing the die.

Stamping techniques may be employed for forming parts with shallow depths, i.e., less than 0.25 inches (in). Water-cooled dies may be used to reduce the time cycle to less than 20 s. This process may be modified by using a rubber block or a fluid-pressured diaphragm as the male form. The formed parts may be reshaped to more rigorous specifications by mechanical reshaping in a process called coining.

Fortunately, ABS is tough and sufficiently ductile to be cold-drawn with a diameter reduction as high as 45 percent. However, the size of the formed part is limited by the dimensions of the sheet, which are usually no more than 14 × 10 feet (ft). The shape of the finished article is also limited by the elongation of the plastic sheet. Some ABS sheets may be stressed as high as 400 to 500 percent over their original area.

The methods of thermoforming have been classified as free forming, drape forming, vacuum forming, plug-assist forming, ridge forming, slip forming, and snap-back forming. These processes may be automated by use of a continuous web, heat tunnels, and conveyors.

In free forming, which is also called free blowing, the sheet is clamped, and vacuum or compressed air is used to blow the sheet into the desired shape. In drape forming, the heated sheet is stretched over

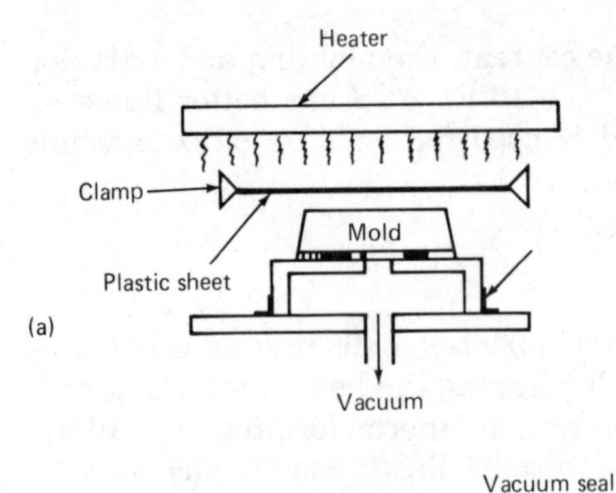

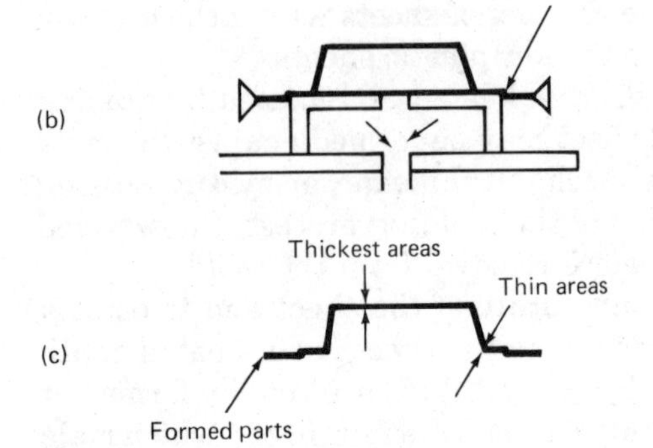

Figure 7.2 Drape forming.

the male mold, and the sheet is allowed to be shaped by the force of gravity or by the application of external pressure or vacuum. The steps in drape forming are illustrated in Fig. 7.2.

As shown in Fig. 7.3, a clamped heated sheet is used in vacuum forming, in which the formed sheet is cooled by contact with the mold surface. Pressure may also be used above the sheet to assist the vacuum-thermoforming process. As shown in Fig. 7.4, a plug is used to force the warm sheet over the male mold in plug-assist forming. This step causes a stretching of the sheet as it moves toward the mold surface. Pressure and/or vacuum are used to transfer the sheet from the surface of the plug to that of the mold.

Since a skeleton frame is used in place of a plug in ridge forming, the sheet has less contact with the plug's surface and thus has no mold marks. Slip forming is an adaptation of metal-drawing techniques, in which the area of the formed sheet is similar to that of the original sheet, regardless of the depth of the draw. This technique is used only with plastics that possess high heat strength.

As shown in Fig. 7.5, the heated sheet is first pulled into a concave shape by vacuum in the technique called vacuum snap-back forming. The male plug is then forced against the sheet, vacuum is applied, and

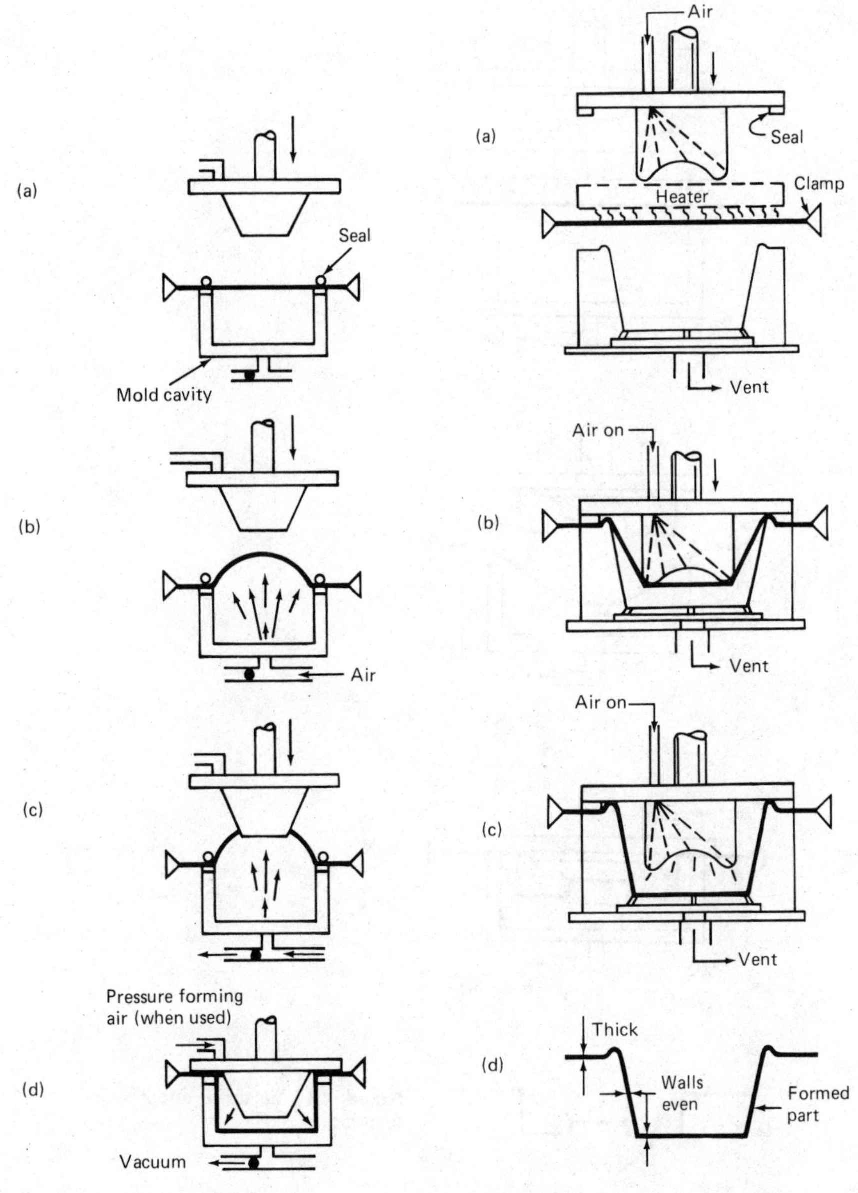

Figure 7.3 Vacuum forming.

Figure 7.4 Plug-assist forming.

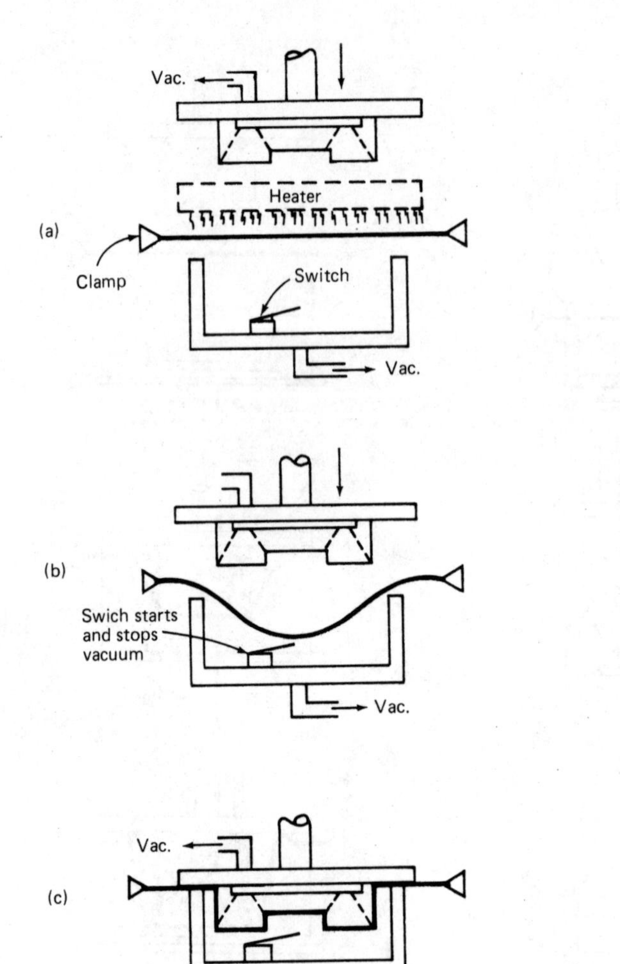

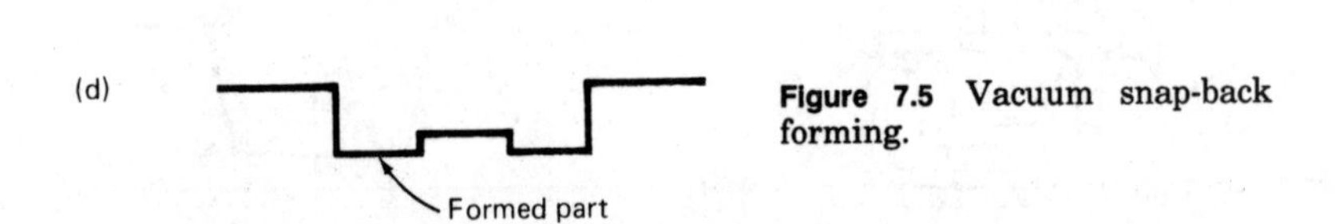

Figure 7.5 Vacuum snap-back forming.

compressed air is used to produce a uniform deep-drawn part, such as luggage. There are many other modifications, such as pressure bubble vacuum snap-back, trapped sheet, contact heat, and pressure forming, which produce formed articles with more uniform thickness.

7.10 Glossary

ABS A terpolymer of acrylonitrile-butadiene and styrene.

acrylonitrile $H_2C{=}CHCN$.

alloy (plastic) A blend of polymers which have properties superior to those of the components.

amorphous Noncrystalline.

Arloy A fiberglass-reinforced SMA-PC alloy.

ASA Acrylic-styrene-acrylonitrile terpolymer.

Barex A copolymer of acrylonitrile and methyl acrylate produced in the presence of NBR.

barrier films Films resistant to gaseous permeation.

bezel A grooved ring which holds a plastic lens in place

Cadon A terpolymer of styrene-maleic anhydride and acrylonitrile and modifications thereof.

carcinogen A cancer-producing substance.

closed die forming Thermoforming of a plastic sheet by the mating of two dies.

coining The reshaping of a thermoformed part.

creep A slow change in physical properties over long time intervals.

Cyclolac ABS.

drape forming Thermoforming of a warmed plastic sheet using the force of gravity to cause it to drape about the male mold.

EPM A copolymer of ethylene and propylene.

extrusion A process in which a heat-softened thermoplastic is forced through the orifice of a die.

fillet A flat surface that separates rounded or angular molded surfaces.

free forming Free blowing, i.e. thermoforming using compressed air.

glass transition temperature The temperature at which a brittle amorphous polymer becomes flexible when heated.

graft copolymer A copolymer in which the pendant groups are actually long chains of a polymer which differs from that making up the principal chain or backbone.

heat deflection temperature The temperature at which a bar deflects a specific distance under a load of 266 lb/in^2.

impact resistance The ability of a material to resist shock loading.

Lopac A modified SAN copolymer.

Luran A SAN.

Lustran An ABS.

Lustron A SAN.

Magnum An ABS.

NBR An elastomeric copolymer of acrylonitrile and butadiene.

olefin An unsaturated hydrocarbon, i.e., a homologue of ethylene.

OSA Olefin-modified SAN.

PAN Polyacrylonitrile: $-\!\!(CH_2CHCN)\!\!-$.

PC Polycarbonate.

PMMA Polymethyl methacrylate.

Polystyrene $-\!\!(CH_2CH(C_6H_5))\!\!-$.

PVC Polyvinyl chloride.

ridge forming A process in which a skeletal frame is used to force the heated sheet toward the male mold.

Rovel A SAN.

SAN Styrene-acrylonitrile copolymer.

silicone A polysiloxane $-\!\!(Si(R_2)O)\!\!-$.

SMA Styrene-maleic anhydride copolymer.

solid-phase forming Forming of a thermoplastic sheet under pressure of moderate temperatures.

Sterem An alloy of SMA-PC.

Styrolux An alloy of SMA-PC.

Taitalac An ABS.

Terpolymer A copolymer produced by the polymerization of three different monomers.

Thermofil A SAN.

thermoform The forming of a heated thermoplastic sheet.

Tyril A SAN.

vacuum snap-back technique A thermoforming technique in which the formed sheet is formed by the use of vacuum and reformed by the use of compressed air.

7.11 References

Boundy, R. H., and R. F. Boyer: *Styrene: Its Polymers, Copolymers and Derivatives,* Reinold Publishing Corporation, New York, 1952.

Brighton, C. A., G. Pritchard, and G. A. Skinner: *Styrene Polymers: Technology and Environmental Aspects,* Applied Science Publishers, Ltd., London, 1979.

Calvert, W. C.: ABS, U.S. Patent 2,908,661, 1959.

Daley, L. E.: ABS, U.S. Patent 2,435,202, 1948.

Frados, J.: *Plastics Engineering Handbook,* Van Nostrand Reinhold Company, Inc., New York, 1976.

Garner, D. P., and G. A. Stahl: *The Effect of Hostile Environments on Coatings and Plastics,* ACS Symposium Series 229, Washington, D.C., 1983.

Irwin, D.: *Modern Plastics Encyclopedia* 64(10A): 286 (1987).

Perry, F. M.: *Acrylonitrile Polymers,* in J. I. Kroschwitz (ed.): *Encyclopedia of Polymer Science and Engineering,* vol. 1, John Wiley & Sons, Inc., Interscience Publishers, New York, 1985.

Seymour, R. B.: *Plastics vs. Corrosives,* John Wiley & Sons, Inc., Interscience Publishers, New York, 1982.

Seymour, R. B., and G. S. Kirshenbaum (eds.): *High Performance Polymers: Their Origin and Development,* Elsevier Science Publishing Company, Inc., New York, 1986.

Seymour, R. B., and J. P. Kispersky: Cadon, U.S. Patent 2,439,227, 1948.

Teach, W. C., and G. C. Kiessling: *Polystyrene,* Reinhold Publishing Corporation, New York, 1960.

Chapter

8

The Acrylics

8.1 Introduction

Since some purists may not include acrylics in their list of engineering polymers, it may be in order to call these polymers strategic materials. Since their volume is much smaller than that of the "big three" thermoplastics, namely, polyolefins, PVC, and PS, and their thermal resistance is less than that of engineering polymers, such as nylon, the acrylics are often overlooked when designers are attempting to solve a material-of-construction problem. Nevertheless, 302,000 t of acrylic polymers was produced in the United States in 1987. The worldwide production was about 680,000 t.

Cast sheets of PMMA were essential for windshields in World War II aircraft, and these sheets continue to be used at an annual rate of 111,000 t in the United States. In addition, almost 90,700 t of PMMA sheet is extruded annually in the United States. Also, as stated in Chap. 6, acrylic elastomers are essential for many solvent- and heat-resistant applications.

Likewise, polyisobutyl acrylate is dissolved in lubricating oil to ensure a high viscosity at the temperature of combustion engines. Copolymers of acrylonitrile, under the trade names Orlon and Acrilan, are widely used as strong fibers, and cyanoacrylates, which polymerize in situ, are excellent adhesives and essential components of modern surgery.

While there are many producers of the "big three" general-purpose polymers, there are only a few producers of acrylics in the United States, namely, Rohm & Haas, Hoechst Celanese, BASF, Du Pont, Union Carbide, and CYRO Industries. The properties of some of these strategic polymers will be discussed in this chapter.

8.2 Acrylic Ester Polymers

Acrylic esters ($H_2C{=}CHCOOR$) have been produced by the hydrolysis of ethylene cyanohydrin ($HOCH_2CH_2CN$) and by the nickel carbonyl [$Ni(CO)_4$] carbonylation of acetylene ($HC{\equiv}CH$). This process, which was one of the many developed by Reppe, has been abandoned in the United States in favor of the catalytic oxidation of propylene, followed by esterification of the acrylic acid produced. Polyacrylic acid is used in aqueous solutions for enhanced oil recovery, oil drilling mud additives, flocculating agents, dental cements, and adhesives. Polyalkyl acrylates are obtained by the free-radical-initiated polymerization of the monomers in bulk, suspension, or emulsion polymerization processes.

Polymethyl acrylate is a tough, rubbery, moderately hard polymer with little tackiness. The rubbery and tacky properties increase as the size of the alcohol used for esterification of the acrylic acid increases. However, because of an attraction between long pendant groups (side-chain crystallization), the polymers become more brittle when the esterifying alcohol contains more than eight carbon atoms (octyl alcohol). Acrylic ester polymers are used as textile finishing agents, pigment binders, paper coatings, leather finishes, caulking compounds, floor polishes, and coatings to prevent insect attack on food. However, most acrylic esters are used to produce copolymers with vinyl acetate and methyl methacrylate.

8.3 Cyanoacrylic Ester Polymers

In 1951, H. Coover discovered that 2-cyanoacrylic esters polymerize rapidly in the absence of an added initiator at room temperature to produce a strong bond between various materials. This discovery was made when he attempted to determine the index of refraction of cyanoacrylate by the use of prisms in an Abbé refractometer. It was not possible to separate the prisms after the film of cyanoacrylate was applied.

Cyanoacrylic adhesives (Krazy Glue) have been used to form bonds with shear strengths in excess of 500 lb/in^2 with acrylic, ABS, phenolic, PC, polysulfone, and PVC. In most cases, it was not possible to break the bond without breaking the plastic. It is not possible to obtain good bonds with polyolefins or polyfluorocarbons, but good bonds are obtained with wood, nylon, PS, polyesters, and polyacetals, as well as with metals. Unfortunately, these bonds deteriorate slowly in the presence of moisture and at temperatures above 71°C (160°F).

8.4 Polyacrylamide

Acrylamide ($H_2C{=}CHCONH_2$) is a white crystalline solid which is readily polymerized by free-radical initiators to produce water-soluble polyacrylamide $[-(H_2CCHCONH_2)-]$. Nylon 3 $[-(CH_2CH_2CONH)-]$ is produced when anionic initiators are used. Polyacrylamide is produced by polymerization of the acrylamide in aqueous solutions.

Random copolymers of acrylamide and vinylpyrrolidone (HE Polymer) have been used commercially for enhanced oil recovery. Graft copolymers of acrylamide and cellulose, starch, and dextran have been produced and proposed for use in enhanced oil recovery. Hydrolysis of graft copolymers of acrylic esters with starch produces a water-absorbent material (Super Slurper). One of the principal uses of polymers and copolymers of acrylamide is as flocculants.

8.5 Flocculants

Small particulates suspended in aqueous media, such as those in sewage and potable-water treatment plants, tend to settle slowly because of their composition and the relation of their size and density to their surface properties. These colloidal particles usually have a charged surface surrounded by a layer of counterions, i.e., a double ionic layer, which prevents settling because of the repulsion of similar charges on the particles, which is the dominant force present.

In contrast, larger suspended particles tend to settle because of the domination of gravitational forces, and the formation of aggregates (flocs) is enhanced by flocculation agents, which aid the precipitation of these suspensoids. Since the charges on the suspended particles are usually negative, they can be neutralized by salts of polymeric acids, such as the sodium salt of polyacrylic acid. Neutral water-soluble polymers, such as polyacrylamide, tend to form "dangling loops" when adsorbed on the surface of suspended particles. Subsequent entanglement of these "dangling loops" results in a "bridging" of particles and precipitation because of their larger size.

Salts of polyacrylic acid, polyacrylamide, and copolymers of acrylic acid and acrylamide are widely used as flocculants. The improvement in flocculation when polyacrylamide is present reduces the settling time and increases the efficiency of water and sewage treatment plants.

8.6 Enhanced Oil Recovery

Oil recovery from wells is not particularly efficient, and as much as 50 percent of the crude oil remains in the rocks in the well. Much of this residual crude can be recovered by injecting a dilute solution of a water-soluble polymer into the pores containing trapped oil. This process is dependent on the interfacial tension between the "thickened water" and the entrapped oil.

It has been observed that hydrolyzed polyacrylamide is not useful for enhanced oil recovery at temperatures above 77°C (190°F). However, copolymers of acrylamide and vinylpyrrolidone are stable, in the presence of brine, at temperatures as high as 121°C (250°F). While this copolymer (HE Polymer) is effective in enhanced oil recovery, the requirement of about 2 lb. per barrel of recovered oil is not economical unless the price of oil is at least $25 a barrel.

8.7 Polyalkyl Methacrylates

Acrylic acid was produced by Redtenbacher, who oxidized acrolein ($H_2C{=}CHCHO$) in 1843, but little was published on polymers of acrylic esters until O. Rohm described these products in his doctoral thesis in 1901. The early emphasis was on polymers of alkyl acrylates, which were used as the inner layer of safety glass in 1933. However, a copolymer of methyl methacrylate and ethyl acrylate was superior to homopolymers of acrylates. Those associated with this application soon recognized that the inner clear polymeric layer had potential use as a glazing material.

The first commercial polymethacrylate sheets were produced by polymerization of the monomer between glass plates (casting) by Rohm & Haas, Du Pont, and ICI in the 1930s under the trade names Plexiglas, Lucite, and Diakon. The arrangement of the ester pendant groups in PMMA $[\!\!-\!\!(CH_2CCH_3COOCH_3)\!\!-\!\!]$ may be all on one side of the chain (isotactic), alternating on each side of the chain (syndiotactic), or random (atactic). The T_g's of these isomers are 209°C (409°F), 160°C (320°F), and 105°C (221°F), respectively. The atactic isomer, produced by free-radical polymerization, is the commercial PMMA.

Because of the increasing size of the pendant groups in esters of higher-molecular-weight alcohols, the T_g's decrease. Thus, while the T_g of atactic PMMA is 105°C (221°F), the T_g of atactic polybutyl methacrylate is −24°C (−11°F). The value of T_g may also be varied by copolymerization of methyl methacrylate with other monomers, such as ethyl acrylate. Thus, a copolymer with any specified T_g value in the range of 6°C (43°F) to 105°C (221°F) may be obtained by the use of appropriate ratios of these monomers.

The T_g of random copolymers may be calculated from the following equation, in which W_1 and W_2 are equal to the weights of the comonomers, and their T_g's are shown in kelvin (K), where K = 0.555(°F − 32) + 273.

$$\frac{1}{T_g} = \frac{W_1}{T_{g_1}} + \frac{W_2}{T_{g_2}}$$

The resistance to aqueous acids and alkalies decreases as the acrylic content in the copolymer increases.

Cast and extruded sheet account for over 50 percent of all PMMA consumed. The end use of the 86,200 t of extruded sheet and moldings produced annually in the United States is as follows: building (55%), industrial (21%), transportation (16%), signs (5%), and consumer use (2.5%).

Sheets which are more readily molded are made from copolymers of styrene and methyl methacrylate. Properties of cast and molded PMMA, as well as styrene–methyl methacrylate copolymer sheets, are shown in Table 8.1.

The tensile strength and percent elongation of PMMA and polymethyl acrylate are 9000 and 1000 lb/in^2, and 4 and 750 percent, respectively. These values for polyethyl acrylate are 10 lb/in^2 and 1800 percent. The values for random copolymers are in between these values and are related to the ratio of each monomer present in the reaction mixture. Simulated marble is made by the in situ polymerization of methyl methacrylate in a mixture with a filler, such as ground limestone. It is customary to add a promoter, such as dimethylaniline, to accelerate this polymerization at ordinary temperatures.

8.8 Optical Properties of Acrylics

PMMA transmits light almost perfectly in the range of 360 to 1000 nanometers (nm), which includes the visible-light range (400 to 700 nm). PMMA sheets do not absorb light, but ultraviolet (UV) absorbers, such as *o*-hydroxybenzophenone, can be added before polymerization to reduce transparency in the UV range (290 to 350 nm). PMMA rod can be used to transmit light around corners, etc.

8.9 Electrical Properties

PMMA has excellent electrical properties and hence can be used in outdoor high-voltage applications. The electrical properties of a PMMA sheet are shown in Table 8.2.

8.10 Glazing Materials

Acrylic sheet is available as colorless sheet, transparent sheet, translucent colored sheet, transparent colored sheet, opaque colored sheet, patterned sheet, double-skinned sheet (with a hollow core), and

TABLE 8.1 Properties of Typical Acrylic Sheets

Property, units*	Cast PMMA	Extruded PMMA	SMA copolymer
Melting point T_m, °F	—	—	—
Glass transition temp. T_g, °F	95	95	100
Processing temp., °F	—	350	350
Molding pressure, 10^3 lb/in^2	—	15	20
Mold shrinkage, 10^{-3} in/in	—	2.5	2
Heat deflection temp. under flexural load of 264 lb/in^2, °F	180	180	210
Maximum resistance to continuous heat, °F	125	125	170
Coefficient of linear expansion, 10^{-6} in/(in · °F)	41	41	40
Compressive strength, 10^3 lb/in^2	15	15	13
Izod impact strength, ft · lb/in of notch	0.3	0.3	0.3
Tensile strength, 10^3 lb/in^2	9	9	10
Flexural strength, 10^3 lb/in^2	15	15	17.5
Percent elongation	5	5	4
Tensile modulus, 10^3 lb/in^2	400	400	450
Flexural modulus, 10^3 lb/in^2	425	425	400
Rockwell hardness	M90	M90	M75
Specific gravity	1.18	1.18	1.11
Percent water absorption	0.3	0.3	0.15
Dielectric constant	140	140	150
Dielectric strength, V/mil	500	500	450
Resistance to chemicals at 75°F:†			
Nonoxidizing acids (20% H_2SO_4)	S	S	S
Oxidizing acids (10% HNO_3)	U	U	U
Aqueous salt solutions (NaCl)	S	S	S
Polar solvents (C_2H_5OH)	S	S	S
Nonpolar solvents (C_6H_6)	U	U	U
Water	S	S	S
Aqueous alkaline solutions (NaOH)	S	S	S

*lb/(in^2 · 0.145) = KPa (kilopascals); ft · lb/(in of notch · 0.0187) = cm · N/cm of notch.
†S = satisfactory; Q = questionable; U = unsatisfactory.

mirrorized sheet. Both 1.25- and 1.75-in-thick bullet-resistant cast acrylic sheets are available in triangular shapes which can be suspended from a tubular aluminum structure for the construction of geodesic domes. The Astrodome in Houston consists of supported PMMA sheets. The original objective of growing grass in this dome was not achieved, and the loss of baseball games by the Astro team was attributed to the glare of the sun. Hence, all panes of PMMA in the Astrodome were coated with pigmented PMMA.

The tensile and flexural strength and tensile modulus of PMMA sheet decrease with an increase in temperature. Some deformation of PMMA and other thermoplastic sheets takes place under load over a

TABLE 8.2 Electrical Properties of 2-in-Thick PMMA Sheet

Dielectric strength, V/mil	500
Dielectric constant	
At 60 Hz	140
At 1000 Hz	132
Power factor, V/mil	
At 60 Hz	0.05
At 100 Hz	0.04
At 1,000,000 Hz	0.02
Loss index (loss factor), V/mil	
At 60 Hz	0.20
At 100 Hz	0.15
At 1,000,000 Hz	0.05
Arc resistance	No tracking
Volume resistivity, ohm-centimeters ($\Omega \cdot$ cm)	5×10^{15}
Surface resistivity, ohms per square (Ω/square)	1.5×10^{17}

period of time. The initial increment of deformation is instantaneous and reversible. Additional deformation is the result of uncoiling and slippage of polymer chains and is, for the most part, irreversible. The change in modulus of elasticity with time and temperature of a PMMA sheet is shown in Table 8.3.

The use of PMMA sheet is not recommended under conditions for which the modulus is less than 100,000 lb/in^2. The effect of a longer time on the modulus can be extrapolated on a logarithmic graph. Scratches have an adverse effect on the flexural strength of PMMA sheets.

Fatigue strength, based on repeated flexural stress, is affected by temperature, relaxation, and loading time. The maximum allowable surface stress at various temperatures is shown in Table 8.4.

Small cracks that are perpendicular to the surface may result from residual stresses developed during thermoforming, contact with solvent vapor, machining, or excess loading. This crazing is irreversible and may precede failure. Since amorphous polymers like glass are

TABLE 8.3 Effect of Time and Temperature on the Modulus of Elasticity (lb/in^2 × 10^3) of PMMA

	Time of loading, days			
Temperature, °F	1	10	100	1000
77	360	310	272	237
104	325	270	222	182
122	315	250	200	160
140	290	220	169	128
158	270	200	145	107
176	240	165	110	—
194	190	110	—	—

TABLE 8.4 Maximum Allowable Surface Stress for PMMA

	Load, $lb/in^2 \times 10^3$	
Temperature, °F	Continuous	Intermittent
77	1.5	3
92	1.3	2
104	1.2	1.5
122	1.1	1.1
140	1.0	1.0
158	0.8	0.8
176	0.6	0.6
194	0.3	0.3

notch-sensitive, impact resistance is determined on notched cantilevered specimens. Hence, the impact resistance is actually much greater than that recorded in Table 8.1 but decreases if cracks are present. Unlike the impact resistance of most other plastics, that of PMMA is not affected to any great extent by changes in temperature. PMMA sheet may be thermoformed using the techniques described in Chap. 7.

Since the coefficient of expansion of PMMA and other plastics is much greater than that of aluminum, steel, or plate glass, allowances must be made for this expansion if the PMMA sheet is used under different temperature conditions. The coefficients of expansion of PMMA, aluminum, steel, and plate glass are 41, 13, 6, and 5×10^{-6} in/(in · °F), respectively.

Moisture also affects the expansion of PMMA and other plastic sheets. Hence, it is recommended that allowances for expansion of 0.06 in/ft of length, over a 28°C (50°F) range, be considered in the design of clear PMMA structures. This value should be doubled for dark-tinted PMMA sheet.

PMMA domes and arches must be designed to be thick enough to resist failure by buckling. PMMA, like most organic thermoplastics, softens, distorts, and melts at elevated temperatures, such as those present in a burning building. Also, PMMA, like many other organic thermoplastics, supports combustion. These inherent properties must be considered in the design of structures. Of course, the flammability can be decreased by the addition of flame retardants, but these additives reduce the clarity of PMMA sheet.

It is important to note that design criteria for plastics in building applications are seldom available to the designer. Hence, the selection of plastic, such as a clear PMMA sheet, should be based on anticipated service. Fortunately, some design data are available from plastic manufacturers, and this information is being expanded as more case history and design data become available.

8.11 Glossary

Abbé refractometer An instrument in which liquid is placed between two prisms in order to measure the index of refraction.

acetylene $HC{\equiv}CH$

acrolein $H_2C{=}CHCHO$

acrylamide $H_2C{=}CHCONH_2$

acrylic A generic term used to describe polymers based on esters of acrylic or methacrylic acid.

acrylic ester $CH_2{=}CHCOOR$

acrylonitrile $CH_2{=}CHCN$

additive Any material added to a polymer, such as a flame retardant.

adsorption The adhesion of a substance to a polymer surface.

aggregate A group of particles held together by weak forces.

alkyl a general term (R) for $H(CH_2)_n$.

amorphous Noncrystalline.

arc resistance The resistance to tracking by high-voltage discharge.

atactic Having the pendant groups of a polymer arranged in random fashion.

"big three" thermoplastics Polyolefins, polyvinyl chloride, and polystyrene.

bridging The entanglement of chains of polymers adsorbed on suspended particles.

casting The formation of a solid by polymerization of a monomer in situ in a container.

catalyst An additive which accelerates the attainment of equilibrium in a chemical reaction.

catalytic oxidation Oxidation in the presence of a catalyst.

colloid Small aggregates of molecules, with a high ratio of surface area to volume, which do not settle out when suspended in a liquid.

copolymer A polymer consisting of two repeating units in the polymer chain.

crazing Defects based on multiple fine cracks on a polymer surface.

creep A dimensional change over a long period of time.

2-cyanoacrylate $CH_2{=}C(CN)COOR$

Diakon A PMMA.

dielectric constant (permittivity) Capacitance of material as dielectric/capacitance of air as dielectric.

dielectric breakdown voltage The voltage between electrodes at which a specimen fails.

dielectric strength The dielectric breakdown voltage per unit thickness.

dimethylaniline $C_6H_5N(CH_3)_2$

dissipation factor The tangent of the dielectric loss angle.

elasticity The reversible recovery of a stretched polymer to its original dimensions.

elongation The fractional increase in length of a stressed specimen.

enhanced oil recovery The recovery of petroleum crude by water flooding after no more oil can be recovered by pumping.

extrusion A process in which an Archimedean screw is used to force a softened thermosplatic through a die.

flocculant An additive which produces coagulation of suspended particles.

flocs Larger particles formed by the addition of a flocculant to small suspended particles.

free radical an electron-deficient molecule (R ·).

free-radical polymerization Addition-type polymerization initiated by a free radical.

glass transition temperature The temperature at which a brittle amorphous solid polymer becomes flexible as the temperature is increased.

glazing A glasslike substance.

graft copolymer A copolymer in which the pendant groups are different polymer chains attached to the main chain of a polymer.

HE Polymer A water-soluble copolymer used for enhanced oil recovery.

hydrolysis Chemical decomposition in the presence of water.

Hz Hertz.

impact resistance Resistance to fracture by shock loading.

isotactic Having pendant groups of a polymer all on one side of the chain.

kelvin A temperature scale in which 0 K is absolute zero.

Krazy Glue A 2-cyanoacrylate.

loops, dangling Portions of polymer chains not adsorbed on suspended particles.

loss factor (loss index) The dielectric constant dissipation factor ratio.

Lucite A PMMA.

modulus of elasticity The stress/strain ratio.

monomer A reactant used to produce a polymer.

notch sensitivity The susceptibility to crack generation in a notched specimen.

nylon A polyamide, such as nylon 6,6.

nylon 3 $+CH_2CH_2CONH_2+$

oil, trapped Residual crude oil not recovered by standard pumping procedures.

pendant group A functional group on a polymer chain.

permittivity The dielectric constant.

Plexiglas A PMMA.

PMMA Polymethyl methacrylate: $+CH_2C(CH_3)(COOCH_3)+$

polyacrylamide $+H_2CCH(CONH_2)+$

polyacrylic acid $+CH_2CH(COOH)+$

polymethyl acrylate $+CH_2CH(COOCH_3)+$

polyolefins A polyhydrocarbon, such as polyethylene: $+CH_2CH_2+$.

polystyrene $+CH_2CH(C_6H_5)+$

polyvinylpyrrolidone $-CH_2-CH-$ (N-substituted 2-pyrrolidone ring: N, =O)

power factor The loss of electric energy in a capacitor when a specific material is used as the dielectric.

PVC Polyvinyl chloride: $+CH_2CHCl+$.

Rohm, O. An early investigator of commercial acrylic polymers.

simulated marble A solid produced by the in situ polymerization of a monomer, such as methyl methacrylate, in the presence of a filler, such as ground limestone.

stress cracking Cracking caused by tensile stresses.

Super Slurper A graft copolymer of acrylic acid and starch.

syndiotactic Having pendant groups alternating on both sides of the polymer chain.

T_g Glass transition temperature.

thermoforming The shaping of a sheet by heat.

thermoplastic A material which may be reversibly softened by heat.

thickened water Water containing dissolved polymers.

translucent Allowing light transmission (usually referring to an opaque specimen).

ultraviolet range 290 to 350 nm (wavelength).

vinyl acetate $CH_2{=}CH(OOCCH_3)$

visible-light range 400 to 700 nm (wavelength).

volume resistivity The electric resistance between opposite faces of a 1-in cube.

8.12 References

Coover, H. W., and J. M. McIntire: "2-Cyanoacrylic Ester Polymers," in J. I. Kroschwitz (ed.): *Encyclopedia of Polymer Science and Engineering,* vol. 1, Interscience Publishers, John Wiley & Sons, Inc., New York, 1985.

Dietz, A. G. H.: *Plastics for Architects and Builders,* The M.I.T. Press, Cambridge, Mass., 1969.

Kine, B. B., and R. W. Novak: "Acrylic and Methacrylic Ester Polymers," in J. I. Kroschwitz (ed.): *Encyclopedia of Polymer Science and Engineering,* vol. 1, Interscience Publishers, John Wiley & Sons, Inc., New York, 1985.

Montella, R.: *Plastics in Architecture,* Marcel Dekker, Inc., New York, 1985.

Miller, E.: *Plastics Products Design Handbook,* Marcel Dekker, Inc., New York, 1981.

Seymour, R. B.: *Plastics vs Corrosives,* Interscience Publishers, John Wiley & Sons, Inc., New York, 1982.

Stachin, J. O.: *Acrylic Plastic Viewpoint for Ocean Engineering Applications,* Naval Undersea Center, San Diego, 1977.

Thomas, W. M., and D. W. Wang: "Acrylic and Methacrylic Acid Polymers," in J. I. Kroschwitz (ed.): *Encyclopedia of Polymer Science and Engineering,* vol. 1, Interscience Publishers, John Wiley & Sons, Inc. New York, 1985.

Chapter

9

Polycarbonates

9.1 Introduction

PCs are amorphous polymers which are used in place of PMMA in applications requiring a clear polymer which is resistant to higher temperatures than PMMA. However, since PC is made by a condensation reaction, it cannot be cast from its reactants, as PMMA can, and hence clear PC sheets must be molded or extruded.

PC was synthesized by Einhorn in 1896, but this clear, tough, heat-resistant polymer was not commercialized until the 1950s, when it was synthesized independently by Schnell and Fox. Polycarbonate has a heat deflection temperature of 125°C (257°F) and an Izod impact resistance of 14 ft · lb/in of notch (depending on the thickness of the test specimen).

Aliphatic polycarbonates [-(-OROCO-)-] have been produced by the phosgenation of diols (HOROH), and aromatic polycarbonates [-(-OAr—OCO-)-] have been produced by the reaction of phosgene ($COCl_2$) with a large number of bisphenols [$(HOC_6H_4)_2C(CH_3)_2$], but the principal PC is produced from BPA.

Fiberglass-reinforced PC can be used continuously at 121°C (250°F). However, some properties of PC sheet may be impaired by moisture, UV radiation, or processing techniques which produce defects near the surface, which can be monitored by FTIR spectroscopy.

There are now three U.S. firms producing PC: General Electric (Lexan), Mobay (Makrolon), and Dow Chemical (Calibre). Over 175,000 t of PC was produced in the United States in 1987.

An optically clear polymer with superior hydrolytic stability, good resistance to stress cracking, and a higher heat deflection temperature has been produced by including phthaloyl isomers in the reaction mixture. The heat deflection temperature is related to the ester content in this aromatic ester carbonate.

PC is being blow-molded for use as compact discs, eyeglass lenses, large water containers, and outdoor lamp shades. Large objects can be fabricated by solvent welding PC using methylene chloride as the solvent.

9.2 Optical Properties of PC

The light transmission through PC is 89 percent. This value compares favorably with that of glass, which has a transmission of 91 percent. PC, like PMMA, is opaque to the long-wave infrared portion of the electromagnetic spectrum but has a high transmittancy to areas of the spectrum with high solar energy. This results in a "greenhouse effect," i.e., solar energy is transmitted through the PC sheet and little reradiation occurs. The U factors of several thicknesses of PMMA, PC, and glass are shown in Table 9.1. The U factor is related to the energy transmitted through a clear sheet. If the energy is not radiated, the temperature below the sheet increases. The U factor is based on an indoor temperature of 23°C (73°F) with an outdoor wind velocity of 15 miles per hour (mi/h).

Special optical-quality PC is commercially available. PC has an index of refraction of 1.58, while the value for PMMA is 1.50. The high value for PC is related to the highly aromatic character of the BPA polycarbonate [$\text{-(}OC_6H_4C(CH_3)_2C_6H_4COO\text{)-}$]. UV radiation, which is absorbed by unstabilized PC, causes yellowing of PC and a reduction in its resistance to impact. A photo-Fries rearrangement takes place when unstabilized PC is exposed to UV radiation. The dihydroxybenzophenones produced absorb in the visible range and cause yellowing. Also, the isopropylidene groups $\text{-(}C(CH_3)_2\text{)-}$ in the bisphenol portion of the repeating units form hydroperoxides which decompose and cause degradation. Thus when used outdoors, PC must contain UV stabilizers. UV-stabilized PC has less tendency to yellow and lose its characteristic impact resistance. Such PC sheets

TABLE 9.1 U Factor for Transparent Sheets, Btu/(h · ft² · °F)*

	Sheet thickness, in		
	0.125	0.175	0.250
PC or PMMA	0.99	0.92	0.85
Glass	1.14	1.13	1.11

*Btu = British thermal unit. These factors are for an indoor temperature of 23°C (73°F) with an outdoor wind velocity of 15 mi/h.

have been used over a period of 10 years with little loss of light transmission or impact resistance.

9.3 Effect of Moisture on PC

Organic carbonates are readily hydrolyzed by water at temperatures above 60°C (140°F) and in the presence of aqueous acids or bases at lower temperatures. Accordingly, it is essential that the PC molding powder be essentially moisture-free (< 0.2% moisture) prior to processing. However, providing the polymer is dry before extrusion or molding, the fabricated articles can be used in aqueous environments, for example, as water bottles, baby bottles, beverage pitchers, wine carafes, and greenhouse glazing. PC dishes and trays have impact strengths that are about 200 times that of glass, which weighs much more than PC. PC articles can be washed in domestic dishwashers.

9.4 Electrical Properties of PC

As shown by the data in Table 9.2, PC has excellent electrical properties. However, as the temperature is increased, the volume resistivity of PC decreases but the dielectric strength increases. The dissipation factor at 60 Hz increases slowly to 0.002 at 141°C (285°F), but increases rapidly at higher temperatures. PC has poor arc resistance.

9.5 PC Sheet

PC sheet, which is the toughest transparent sheet available, is produced in thicknesses up to 0.5 in by extrusion through a slit die. Thicker sections can be fabricated by lamination using methylene

TABLE 9.2 Electrical Properties of PC

Dielectric strength, V/mil	380
Dielectric constant	
At 60 Hz	3.2
At 1,000,000 Hz	3.0
Power factor	
At 60 Hz	0.0009
At 1,000,000 Hz	0.010
Volume resistivity, Ω · cm	$>10^{16}$
Arc resistance, s	
Stainless-steel electrodes	10
Tungsten electrodes	120

TABLE 9.3 Properties of Typical 0.125-in-thick PC Sheet

Property, units*	
Melting point T_m, °F	—
Glass transition temp. T_g, °F	150
Processing temp., °F	550
Molding pressure, 10^3 lb/in^2	15
Mold shrinkage, 10^{-3} in/in	5
Heat deflection temp. under flexural load of 264 lb/in^2, °F	270
Maximum resistance to continuous heat, °F	250
Coefficient of linear expansion, 10^{-6} in/(in · °F)	38
Compressive strength, 10^3 lb/in^2	12.5
Izod impact strength, ft · lb/in of notch	14
Tensile strength, 10^3 lb/in^2	9
Flexural strength, 10^3 lb/in^2	13.5
Percent elongation	100
Tensile modulus, 10^3 lb/in^2	345
Flexural modulus, 10^3 lb/in^2	340
Rockwell hardness	M92
Specific gravity	1.2
Percent water absorption	0.5
Dielectric constant	3.2
Dielectric strength, V/mil	380
Resistance to chemicals at 75°F:†	
Nonoxidizing acids (20% H_2SO_4)	Q
Oxidizing acids (10% HNO_3)	U
Aqueous salt solutions (NaCl)	S
Polar solvents (C_2H_5OH)	S
Nonpolar solvents (C_6H_6)	U
Water	S
Aqueous alkaline solutions (NaOH)	U

*lb/(in^2 · 0.145) = KPa (kilopascals); ft · lb/(in of notch · 0.0187) = cm · N/cm of notch.

†S = satisfactory; Q = questionable; U = unsatisfactory.

chloride as the solvent. The physical properties of a 0.125-in-thick PC sheet are shown in Table 9.3.

9.6 Effect of Variables on Physical Properties of PC

Since the presence of hot moisture causes a reduction in the molecular weight of PC, it is essential that PC be dried before molding or extrusion. Since exposure of PC sheet to solvent vapor causes crazing, such exposure must be avoided. Machining and polishing may promote crazing, and crazing may also occur when PC sheet is used at greater than recommended loads. The maximum allowable loading stress for PC at various temperatures is shown in Table 9.4.

TABLE 9.4 Maximum Allowable Loading Stress for PC

	Loading, lb/in² × 10³	
Temperature, °F	Continuous	Intermittent
- 65	4	5
0	2.3	4.2
73	2	4
130	1.6	3.5
160	1	3.2
200	0.5	3
250	0	2.5

9.7 Effect of Temperature on PC

PC sheets require a higher temperature than PMMA or ABS for forming, but the techniques described in Chap. 7 are satisfactory for thermoforming PC. The coefficient of linear thermal expansion of PC is 38×10^{-6} in/(in · °F), which is similar to that of PMMA. However, since the effect of moisture on expansion and contraction of PC is negligible, an allowance of 0.03 in/ft of length is used in design instead of the 0.06 in/ft used for PMMA. In both cases, the values must be doubled when dark-tinted sheets are used.

The maximum continuous-use temperature (MCUT), i.e., the temperature at which physical strength decreases to 50 percent of its original value in 11.4 years, is 104°C (220°F). The CUT for short-term deformation is 132°C (270°F). Since PC is thermooxidized at temperatures above 299°C (570°F), about 0.5 percent of stabilizers, such as phosphites or silicones, is added to the molding compound.

9.8 Flammability

PC has an oxygen index (OI) of 26 and is listed as V-2 by Underwriters' Laboratory (UL). Hence, PC without flame retardants is satisfactory for many nonelectrical applications. According to UL flammability ratings, V-O is satisfactory and V-5 is unsatisfactory. Transparent flame-resistant PC can be produced by using tetrabromobisphenol A as a partial replacement for BPA in the reactants. Such formulations are listed as V-0 by UL. PC containing salts of flame-retardant additives such as sodium 2,4,5-trichlorobenzene sulfonate as flame retardants are transparent and listed as V-0 by UL. The addition of PTFE reduces dripping from burning PC, but this additive also reduces transparency.

9.9 PC Foams

PC foams are readily produced by chemical blowing agents (CBA), such as 5-phenyltetrazole, or physical blowing agents (PBA), such as nitrogen. It is customary to incorporate a small amount (5%) of fiberglass to serve as a nucleating agent to ensure a homogeneous cellular structure. The properties of a typical PC foam are shown in Table 9.5.

9.10 Reinforced PC

The properties of PC reinforced with 30 percent fiberglass are also shown in Table 9.5.

9.11 Polyester Carbonates

Polyester carbonates, produced by the reaction of $COCl_2$ with a mixture of BPA and isophthaloyl chloride, have higher heat deflection temperature values than PC. The properties of a polyester carbonate made with equal amounts of BPA and isophthaloyl chloride are shown in Table 9.5.

9.12 Reinforced PC

The rigidity, hardness, and dimensional stability are increased and the coefficient of expansion is decreased when PC is reinforced with fiberglass or graphite. The mechanical properties of these composites are improved when coupling agents, such as organosilanes, are added to fiberglass-reinforced PC. The properties of reinforced PC are shown in Table 9.5.

9.13 PC Blends

Polypivalolactone is immiscible with PC, but many other polyesters, such as PET (Xenoy) and PBT (Macroblend) are completely miscible with PC. Other commercial blends are PC-polyethylene (Lexan), PC-polyurethane (Texin), PC–styrene–maleic anhydride copolymer (Arloy), PC-nylon (Dexarb), ABS-PC (Bayblend, Proloy, Pluse), and ASA-PC (Tarblend).

The overall properties P of blends or alloys are equal to the sum of the properties of the components, modified by their concentrations, plus a level of synergistic parameter, I, multiplied by the component

TABLE 9.5 Properties of Typical Polycarbonates

Property, units*	PC	PC foam	PC-polyester	PC + 30% fiberglass	PC + 40% graphite
Melting point T_m, °F	—	—	—	—	—
Glass transition temp. T_g, °F	300	—	175	150	150
Processing temp., °F	550	—	650	600	600
Molding pressure, 10^3 lb/in^2	15	--	15	20	15
Mold shrinkage, 10^{-3} in/in	3.5	—	2	1	1.5
Heat deflection temp. under flexural load of 264 lb/in^2, °F	270	260	310	300	290
Maximum resistance to continuous heat, °F	250	240	300	280	275
Coefficient of linear expansion, 10^{-6} in/(in · °F)	70	—	40	10	5
Compressive strength, 10^3 lb/in^2	12.5	—	11.5	19	23
Izod impact strength, ft · lb/in of notch	16	6.5	8	2.5	2
Tensile strength, 10^3 lb/in^2	9.5	5	10.5	19.5	22
Flexural strength, 10^3 lb/in^2	13.5	—	13.5	24	33
Percent elongation	110	5	90	3	3
Tensile modulus, 10^3 lb/in^2	345	290	250	130	1500
Flexural modulus, 10^3 lb/in^2	340	10	300	960	1200
Rockwell hardness	M70	—	M85	M92	R118
Specific gravity	1.2	—	1.2	1.4	1.32
Percent water absorption	0.15	—	0.2	0.2	0.05
Dielectric constant	3.2	—	3	3	—
Dielectric strength, V/mil	380	—	510	475	—
Resistance to chemicals at 75°F:†					
Nonoxidizing acids (20% H_2SO_4)	Q	Q	Q	Q	Q
Oxidizing acids (10% HNO_3)	U	U	U	U	U
Aqueous salt solutions (NaCl)	S	S	S	S	S
Polar solvents (C_2H_5OH)	S	S	S	S	S
Nonpolar solvents (C_6H_6)	U	U	U	U	U
Water	S	S	S	S	S
Aqueous alkaline solutions (NaOH)	U	U	Q	U	Q

*lb/(in^2 · 0.145) = KPa (kilopascals); ft · lb/(in of notch · 0.0187) = cm · N/cm of notch.
†S = satisfactory; Q = questionable; U = unsatisfactory.

properties P_1P_2. In polymer blends, the value of I approaches zero, i.e., they are nonsynergistic. The value of I is greater than zero for alloys and synergistic mixtures. The equation for calculating the properties of blends or mixtures is

$$P = P_1C_1 = P_2C_2 + IP_1P_2$$

Over 18,000 t of PC blends is consumed annually worldwide.

9.14 Glossary

ABS Acrylonitrile-butadiene-styrene terpolymer.

alloy A synergistic blend, i.e., one in which the overall properties are better than those predictable from the law of mixtures.

Ar An aromatic group.

Bisphenol A $HOC_6H_4C(CH_3)_2C_6H_4OH$

blends Compatible or incompatible mixtures of polymers.

BTU British thermal unit; energy required to increase the temperature of 1 lb of water by 1°F.

Calibre A PC.

carbonic acid HOCOOH

crazing The formation of small surface cracks in the presence of solvent vapors, machining, or stress.

dihydroxybenzophenone

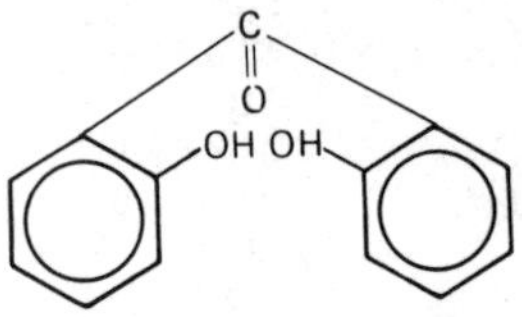

FTIR spectroscopy Fourier transform infrared spectroscopy.

greenhouse effect The increase in temperature (solar heat gain) when energy from the sun passes through a transparent barrier and only a small portion of this energy is reradiated out through the barrier.

I A synergistic parameter.

index of refraction The ratio of the phase velocity of light in a vacuum to that in a specified medium, such as a polymer.

Lexan A PC.

Makrolon A PC.

MCUT The maximum continuous use temperature, i.e., the temperature at which 50 percent of the physical strength of a material is lost after 11.4 years of service.

OI Oxygen index.

oxygen index The minimum oxygen concentration, in an oxygen-nitrogen mixture, which permits downward flame propagation of a candle-type flame.

P The overall properties of a blend.

phosgene $COCl_2$

photo-Fries rearrangement A UV-energized rearrangement in which an ortho-hydroxyl group is formed, as in phenyl salicylate, and these structures are rearranged to form a 2,2′-dihydroxybenzophenone structure.

polycarbonate (PC) $-\!\!\left[O-C_6H_4-C(CH_3)_2-C_6H_4-C(=O)-O \right]\!\!-$

polyester carbonate A copolymer of an aromatic ester and a carbonate.

polymethyl methacrylate (PMMA) $-\!\!\left[CH_2C(CH_3)COOCH_3 \right]\!\!-$

PTFE Teflon.

R An aliphatic group.

salicylic acid $C_6H_4(OH)COOH$

solvent welding The joining of two thermoplastic parts by the addition of solvent.

synergistic parameter *I*: a parameter in which the magnitude is related to the improvement in physical properties of alloys.

U factor A factor which is related inversely to transmitted energy through a clear plastic or glass sheet.

9.15 References

Carhart, R. O.: "Polycarbonates," Chapter 3 in *Engineering Thermoplastics,* Marcel Dekker, Inc., New York, 1985.

Fox, D., and W. Christopher: *Polycarbonates,* Reinhold Publishing Corporation, New York, 1962.

Freitag, D., V. Grigo, P. R. Muller, and W. Nouvertue: "Polycarbonates," in J. I. Kroschwitz (ed.): *Encyclopedia of Polymer Science and Engineering,* vol. 11, Interscience Publishers, John Wiley & Sons, Inc. New York, 1988.

Schnell, H.: *Chemistry and Physics of Polycarbonates,* Interscience Publishers, John Wiley & Sons, Inc. New York, 1964.

Seymour, R. B.: *Polymers for Engineering Applications,* ASM International, Metals Park, Ohio, 1987.

Seymour, R. B., and C. E. Carraher: *Structure-Property Relationships in Polymers,* Plenum Press, Plenum Publishing Corporation, New York, 1984.

Chapter

10

Acetals

10.1 Introduction

POM, like PC, is one of the "big five" engineering polymers. The annual U.S. and worldwide production of POM is 50,000 t and 227,000 t, respectively. This polymer, which is produced when formalin (37% aqueous formaldehyde solution) is kept for long periods of time at room temperature, was investigated by Butlerov in 1859 and reexamined by Staudinger in 1922. It is a tribute to the ingenuity of Staudinger that he was able to write the correct formula for POM, i.e., $-(CH_2-O)-$.

Because of its thermal instability, the commercialization of this important polymer was delayed until R. McDonald of Du Pont capped (esterified) the terminal hydroxyl groups in POM in 1964. Subsequently, in 1967, C. Walling of Celanese also obtained thermally stable POM by making random copolymers of formaldehyde (H_2CO) and other related monomers, such as ethylene oxide

$$(H_2C - CH_2) \text{ with } O \text{ bridging (epoxide ring)}$$

The trade names of the Du Pont and Celanese polymers are Delrin and Celcon, but Celanese also uses the trade names Kemetal and Duracon in Europe and in the Orient, respectively. BASF uses the trade name Ultraform, and Asahi and Mitsubishi use the names Tenac and Iupital for POM.

Acetal homopolymers and copolymers are the stiffest and strongest of the unreinforced engineering polymers. They have excellent load-bearing capacity and are resistant to stress cracking and to wear. They are readily injection-molded and -extruded, but both the homopolymer and the copolymer decompose to produce formaldehyde at temperatures above 218°C (425°F). Because of the ester end groups, the stabilized homopolymer is less resistant to hydrolytic degradation. Both types have fair resistance to organic solvents but are unstable in strong acids, i.e., at a pH of <4.

10.2 Extrusion of POM

POM is a crystalline polymer which possesses many performance characteristics that are similar to those of metals. Like aluminum and most thermoplastics, it may be readily extruded providing the extrusion compound is dry.

As shown in Fig. 10.1, the granulated POM is placed in the hopper *c* where it falls to a rotating archimedean screw at *d* which transports it through zone *a* and then to the metering zone *b*. The polymer melt is pumped via breaker plates and screens through the nozzle *f* and then through the die. The die may be circular for extruded rod, pipe, wire coating, or blown film, or it may have any other desired shapes for the profile extrudate. It is essential that POM not be heated above 249°C (480°F) and that its residence time in the extruder barrel be less than 8 min. The nozzle should be heated at about 204 to 216°C (400 to 420°F), and this temperature should be monitored at the start and at the end of the extrusion run.

10.3 Injection Molding of POM

Unlike the thermosets which were described in Chaps. 3, 4, and 5, which are usually fabricated by labor-intensive, slow compression-molding techniques, molded articles of POM and other thermoplastics may be produced rapidly by the injection-molding process. Cross sec-

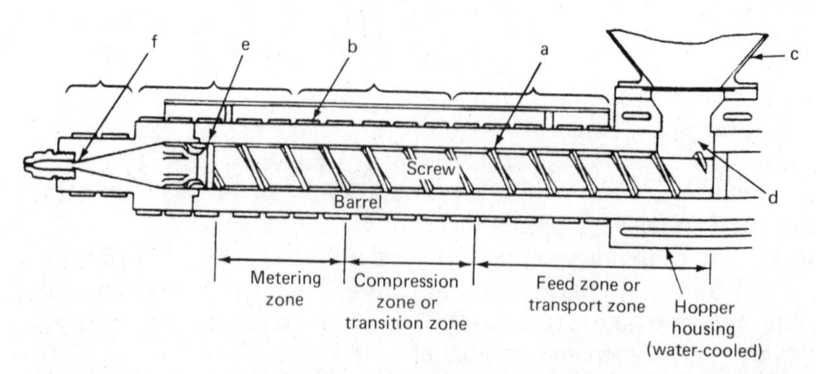

Figure 10.1 Sketch details of screw and extruder zones. (*From* "Modern Plastics Technology" *by R. Seymour, 1975, Reston Publishing Company, Reston, Virginia. Used with permission of Reston Publishing Company.*)

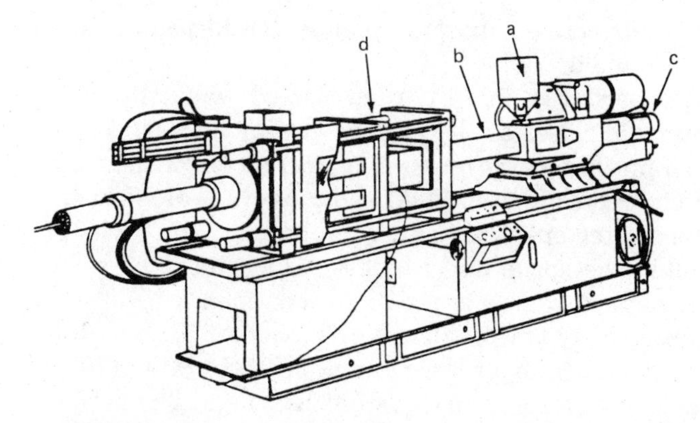

Figure 10.2 Cross section of an injection-molding press. (*From* "Modern Plastics Technology" *by R. Seymour, 1975, Reston Publishing Company, Reston, Virginia. Used with permission of Reston Publishing Company.*)

tions of an injection-molding press and an injection mold are shown in Figs. 10.2 and 10.3. In the molding process, a measured amount of granulated POM molding compound is conveyed from the hopper *a* to a heated cylinder *b*, where it is plasticated (softened) and forced by a

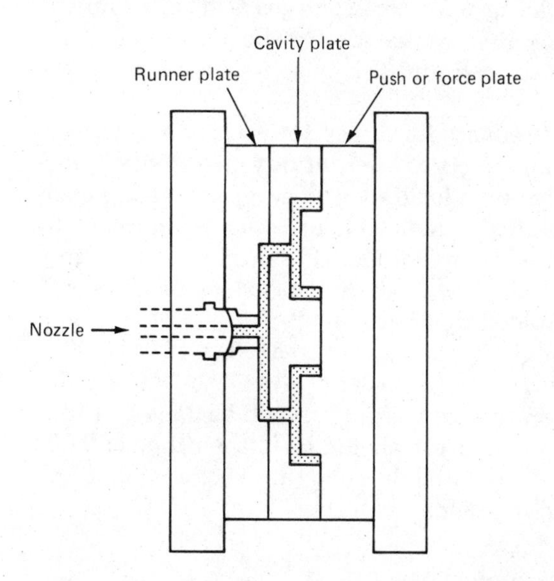

Figure 10.3 Injection mold in closed position. (*Adapted from* "Modern Plastics Technology" *by R. Seymour, 1975, Reston Publishing Company, Reston, Virginia. Used with permission of Reston Publishing Company.*)

reciprocating plunger or screw *c* past a spreader or torpedo into a cooled closed two-piece mold.

The cooled, molded part (Fig. 10.3) is then ejected when the mold opens, and then the mold closes as the plunger begins its next forced stroke in the cyclical mechanical process. A typical mold consists of a sprue, or stock, which accepts molten POM and conveys it along openings (runners) to constricted openings (gates), and then to the cavity of the closed, split mold. The sprue and runners, which must be separated from the molded part, can be ground and remolded.

Since POM has a relatively low coefficient of friction (0.33), which is essentially constant over a temperature range of 21 to 93°C (70 to 200°F), it can be molded to form wear-resistant bearings. The performance of bearings may be predicted from the "*PV* factor," which has been used to determine the utility of sintered metal bearings.

The heat generated in a POM bearing is dependent on the load or pressure on the bearing, *P*, and the surface velocity *V*. The *PV* limits can be increased as much as 500 percent by lubrication. The *P* values are equal to the load/projected area, and *V* is equal to $\pi D_S/12$, where π is expressed as revolutions per min and D_S is the diameter of the shaft used for bushings or journal bearings. The recommended limiting *PV* for POM is 4000.

POM is also used for molding wear-resistant gears. The low modulus of elasticity of POM permits greater tooth deflection under load, since several teeth share the load and this sharing increases as the load increases.

The low modulus of POM reduces its susceptibility to stress concentration. Stress concentration effects are essentially negligible. There is also a reduction in the dynamic loading of POM gears during operation. The dimensional stability of molded gears may be improved by annealing POM. However, while moldings with thin-walled sections are improved by annealing, thick-walled sections are essentially self-annealing if the POM is molded in a hot mold.

It is recommended that high mold temperatures [93 to 121°C (200 to 250°F)] and minimum pressures in the order of 14,000 to 18,000 lb/in^2 be used for molding POM gears. The molding should be done at 193 to 216°C (380 to 420°F), and the gate size should be in the range of 0.003 to 0.010 in. The cycle time should be 20 s for thin sections (0.06 in), and as long as 75 s for thick sections (0.5 in).

10.4 Machining and Buffing of POM

Prototypes of POM can be fabricated by machining extruded rod or molded slab stock. Since POM and many thermoplastics are notch-sensitive, one should use sharp tools with adequate clearance and adequate cooling. POM may be milled, sawed, drilled, turned, threaded, tapped, or punched.

A wheel should be used for buffing and polishing POM. Coarse files should be used for filing, and standard tools, with some modification, should be used for shaping POM.

10.5 Properties of POM

POM has excellent abrasion resistance, excellent dimensional stability, and a low coefficient of friction against metals. The properties of POM as a bearing material are enhanced by the incorporation of PTFE (Teflon) fibers. This composite can be molded on screw-type injection-molding machines. More detailed information on a typical POM plastic is shown in Table 10.1.

10.6 Modified POM

An alloy of POM and an elastomer, which is commercially available (Delrin 100ST, Duraloy 1000), has outstanding toughness but fails in a ductile mode when tested for impact resistance. A 10-fold increase in the modulus of POM is produced by a 20-fold draw in length. POM that is more resistant to oxidation and does not release free formaldehyde is produced by adding stabilizers, such as hindered phenols or amidines.

The high degree of shrinkage of POM during injection molding is reduced by the addition of mineral fillers. Fillers are usually more effective in upgrading crystalline polymers, such as POM, than in upgrading amorphous polymers. UL assigns a 74 to 99°C (165 to 210°F) continuous-use temperature for POM. POM has a higher flexural modulus but lower tensile strength than unfilled POM.

10.7 Electrical Properties of POM

As shown by the data in Table 10.2, POM, like many other thermoplastics, has excellent electrical properties, including a constancy of dielectric constant and low loss in the useful operating range of −40 to 50°C (−40 to 122°F).

10.8 Flammability of POM

POM retains its good mechanical properties at elevated temperatures, but 0.125-in-thick neat POM burns at a rate of 1.1 in/min if ignited by

TABLE 10.1 Properties of Typical POM Plastics

Property, units*	Homopolymer (Delrin)	Copolymer (Celcon)	Impact-modified homopolymer	Impact-modified copolymer	25% fiberglass-reinforced homopolymer
Melting point T_m, °F	350	340	350	340	340
Glass transition temp. T_g, °F	—	—	—	—	—
Processing temp., °F	425	400	400	380	455
Molding pressure, 10^3 lb/in^2	15	12	10	10	15
Mold shrinkage, 10^{-3} in/in	20	20	15	17	8
Heat deflection temp. under flexural load of 264 lb/in^2, °F	255	220	200	170	320
Maximum resistance to continuous heat, °F	240	200	175	150	270
Coefficient of linear expansion, 10^{-6} in/(in · °F)	60	40	70	65	15
Compressive strength, 10^3 lb/in^2	17	16	—	—	17
Izod impact strength, ft · lb/in of notch	1.8	1.2	6	6	1.4
Tensile strength, 10^3 lb/in^2	10	9	7	6	17
Flexural strength, 10^3 lb/in^2	14	13	—	—	25
Percent elongation	50	60	150	110	2.5
Tensile modulus, 10^3 lb/in^2	520	450	275	285	1300
Flexural modulus, 10^3 lb/in^2	400	400	275	200	1100
Rockwell hardness	M92	M85	M70	M55	M79
Specific gravity	1.4	1.4	1.4	1.35	1.6
Percent water absorption	0.3	0.2	0.3	0.3	1.0
Dielectric constant	3.7	3.7	3.7	3.7	3.7
Dielectric strength, V/mil	500	500	450	450	520
Resistance to chemicals at 75°F:†					
Nonoxidizing acids (20% H_2SO_4)	Q	Q	Q	Q	Q
Oxidizing acids (10% HNO_3)	U	U	U	U	U
Aqueous salt solutions (NaCl)	S	S	S	S	S
Polar solvents (C_2H_5OH)	S	S	S	S	S
Nonpolar solvents (C_6H_6)	Q	Q	Q	Q	Q
Water	S	S	S	S	S
Aqueous alkaline solutions (NaOH)	Q	S	Q	S	Q

*lb/(in^2 · 0.145) = KPa (kilopascals); ft · lb/(in of notch · 0.0187) = cm · N/cm of notch.
†S = satisfactory; Q = questionable; U = unsatisfactory.

TABLE 10.2 Electrical Properties of POM

	Homopolymer (Delrin)	Copolymer (Celcon M90)
Dielectric constant	3.7	3.7
Dielectric strength, V/mil	500	500
Dissipation factor		
At 100 Hz	1×10^{-3}	1×10^{-3}
At 1,000,000 Hz	5×10^{-3}	6×10^{-3}
Arc resistance, s	220	240
Comparative tracking index	600+	600+
Volume resistivity, Ω · cm	10×10^{14}	1×10^{14}
Surface resistivity, Ω · cm	1×10^{15}	13×10^{15}

a flame. UL rates POM in its 94 test as HB, i.e., no burn in its horizontal test. The inherent flammability of POM can be decreased by the addition of flame retardants.

10.9 Effect of Water on POM

Both the homopolymer and the copolymer have been used successfully in many applications in hot-air [93°C (200°F)] and hot-water [82°C (180°F)] environments. The UL thermal index (mechanical with impact) ratings for the polymer and the copolymer are 85°C (185°F) and 90°C (194°F), respectively. The UL electrical rating is a bit higher, i.e., 105°C (221°F) and is identical for both types of POM. POMs (acetals) are covered by several local plumbing codes, and by Federal Specification LP-392a. The Society of the Plastics Industry (SPI) recommends a hydrostatic design stress (RHDS) of 1000 lb/in^2 at 23°C (73°F) for POM fittings.

10.10 Chemical Properties of POM

As indicated by the resistance data in Table 6.1, POM is not suitable for use at 23°C (73°F) in the presence of oxidizing acids, such as 10% nitric acid, or in the presence of 5% sodium hypochlorite (NaOCl). However, the copolymer is more resistant to NaOCl than the homopolymer is. Neither is suitable for use in 10% hydrochloric acid or 30% sulfuric acid or in any acid at a pH less than 4.0. The homopolymer should not be used in the presence of 10% aqueous ammonia ($NH_{3_{aq}}$) or 10% sodium hydroxide (NaOH), but the copolymer is somewhat more resistant to these alkaline environments.

Both homopolymer and copolymer can be used in the presence of 5% acetic acid, acetone, benzene, carbon tetrachloride, ethyl acetate, ethyl alcohol, and motor oil. The homopolymer is unsatisfactory for use in the presence of unleaded gasoline. The copolymer is satisfactory in this environment.

POM is unchanged after 1 year in chlorine-free hot-water [82°C (180°F)] service. One of the greatest differences between the polymer and the copolymer in environmental stability is in hot aqueous alkaline solutions, which attack the homopolymer but do not attack the copolymer.

10.11 Resistance of POM to Creep and Fatigue Stress

In general, POM is resistant to creep, i.e., cold flow, which involves a dimensional change under load over a long period of time. These crystalline polymers have unusually good resistance to tensile and flexural fatigue stress. The fatigue endurance limit (FEL) value, i.e., the stress at which a specimen fails in 10^7 cycles, is 3300 to 5000 lb/in^2 for POM at 23°C (73°F) and is somewhat lower at higher temperatures.

10.12 Health and Safety Factors

It had been recognized for several decades that when heated, POM produces formaldehyde. This objection was overcome by capping the end groups with acetyl (H_3CCO—) groups or by copolymerization with monomers such as ethylene oxide. Actually, each copolymer molecule thermally decomposes to produce a few molecules of formaldehyde until the terminal groups of the residue are no longer methylol groups (—CH_2OH). In any case, even the stabilized POM yields formaldehyde when heated to temperatures above 218°C (425°F). Hence, good ventilation must be provided in processing and fabricating plants. The human nose can detect 0.1 to 0.5 parts per million (ppm) of formaldehyde, and commercial devices are available for quantitative detection. Badge-style detectors of formaldehyde can be worn by those exposed to formaldehyde fumes.

Formaldehyde has been used without severe problems for centuries in embalming formulations and for the preservation of biological specimens. Nevertheless, large doses of formaldehyde in animals may be carcinogenic, and proper precautions should be taken when small concentrations are present in the atmosphere. It should be noted that lachrymation, coughing, nasal irritation, and sometimes nausea are observed when the concentrations of formaldehyde are greater than 3 ppm. The threshold limit value (TLV) is 1 ppm of formaldehyde, but this value may be lowered as more data on exposure of laboratory animals become available.

10.13 Glossary

acetal polymera POM: $-(CH_2O)-$

acetyl group H_3CCO-

acid, oxidizing Nitric acid, chromic acid, concentrated sulfuric acid.

alloy A polymer blend with superior properties.

ammonia, aqueous A water solution of ammonia (NH_3).

annealing A heating process which relieves stress in a fabricated polymer article.

Archimedean screw A rotating rod-threaded screw which transports a polymer through a barrel.

big five engineering polymers nylon, aromatic polyesters, acetals, polycarbonates, and polyphenylene oxide.

breaker plate A perforated plate located before the nozzle which supports the screens in an extruder or injection-molding machine.

capping The reaction of end groups (terminal groups) with appropriate reaction.

carcinogenic Cancer-producing.

Celcon A POM.

copolymer A polymer with more than one repeating unit.

creep Cold flow, long-time dimensional change, under load.

Delrin A POM.

Duracon A POM.

ester group $-COOR$

ethylene oxide H_2C-CH_2 (bridged by O)

extrusion The process of forcing a heat-softened thermoplastic through a die and cooling the emerging profile.

FEL Fatigue endurance limit.

formaldehyde H_2CO

formalin An aqueous solution of formaldehyde.

friction, coefficient A measure of the resistance of a polymer to slide over another surface.

gate The orifice through which the molten polymer enters the mold cavity in the injection-molding process.

homopolymer A polymer consisting of one repeating unit.

hopper A container for supplying granulated polymer in an extruder or injection-molding press.

hydrolytic degradation Decomposition in the presence of water.

injection molding A process in which measured amounts of a heat-softened polymer are forced into a cooled, closed split-mold cavity and ejected from the mold after solidification.

Iupital A POM.

Kemetal A POM.

methylol group $—CH_2OH$

modulus The stress/strain ratio.

notch-sensitivity The tendency for failure of a notched glasslike solid.

nozzle A hollow cored metal nose which is designed to serve as a seal between the heated cylinder and the die in an extruder.

pH Log $1/[H_3O^+]$, a scale of 0 to 14 which describes decreasing acidity (increased alkalinity, basicity) of an aqueous solution. The neutral point is 7, the pH of aqueous sodium hydroxide (NaOH) is about 14, and that of hydrochloric acid is about 1.

phenol, hindered A stabilizer in which the substituents are present in positions adjacent to the hydroxyl group (OH) in phenol.

plastication Softening by heating.

POM Polyoxymethylene: $-(CH_2O)-$.

ppm Parts per million.

PTFE Polytetrafluoroethylene: $-(CF_2—CF_2)-$.

***PV* factor** A factor based on the product of load *P* and surface velocity *V* which is used to predict the utility of a bearing material.

runner The channel which connects the sprue and the gate in a closed injection mold.

sintering A pressurized process in which particles are bonded together, without melting.

sodium hypochlorite Bleach, NaOCl.

sprue A vertical channel through which the softened polymer passes to the closed injection mold. This term is also applied to the polymer, which remains in this channel.

Staudinger A pioneer in polymer science.

thermoplastic A polymer which can be reversibly softened by heating.

thermoset A cross-linked or network polymer which cannot be softened by heat.

torpedo A metal block which spreads the softened polymer in an injection-molding machine so that thin layers will be more uniformly heated.

UL Underwriters' Laboratory.

Ultraform A POM.

10.14 References

Dolce, T. J., and J. A. Grates: "Acetal Resins," in J. I. Kroschwitz (ed.): *Encyclopedia of Polymer Science and Engineering,* vol. 1, Interscience Publishers, John Wiley & Sons, Inc. New York, 1985.

Persak, K. J., R. A. Flemming, T. J. Dolce, and F. B. McAndrew: in *High Performance Polymers: Their Origin and Development,* Elsevier Science Publishing Co., Inc., New York, 1986.

Serle, A. G., "Acetals," Chapter 1 in *Engineering Thermoplastics,* Marcel Dekker, Inc., New York, 1985.

Seymour, R. B.: *Modern Plastics Technology,* Reston Publishing Co., Inc., Reston, Va., 1975.

Sittig, M.: *Polyacetal Resins,* Gulf Publishing Company, Houston, 1963.

Chapter

11

Thermoplastic Aromatic Polyesters

11.1 Introduction

Polyesters were investigated by Berzelius in 1847, and thermoset polyesters were introduced in the early years of the twentieth century as Glyptals by Smith, alkyds by R. Kienle, and unsaturated polyesters by Ellis. However, the investigation of the thermoplastic polyesters was delayed when W. H. Carothers concluded that the melting point of these polymers is below the temperature used in pressing clothes with a hot iron.

He also developed the well-known Carothers' equation, which shows that high purity of the diols and dicarboxylite acids is essential for the production of high-molecular-weight polymers. In the following equation, $\overline{DP}$ is the average degree of polymerization, i.e., the number of repeating units in the polymer chain, and p is the extent of reaction. Simple calculations would show that a $\overline{DP}$ of at least 50 cannot be obtained unless p is equal to at least 0.98. Fiber-forming and moldable aliphatic polyesters must have a DP of at least 50.

$$\overline{DP} = \frac{1}{1 - p}$$

Carothers' equation

Because of the flexibility of the sequence of methylene (CH_2) groups in the chains of aliphatic polyesters, they are characterized by low melting points. This problem was solved in 1941 by J. R. Whinfield and J. T. Dickson, who used a dicarboxylic acid, that is, terephthalic acid in place of the aliphatic dicarboxylic acids, such as adipic acid [$HOOC(CH_2)_4COOH$], which has been used as reactants by Carothers. The original aromatic polyesters were used as fibers, namely, Terylene,

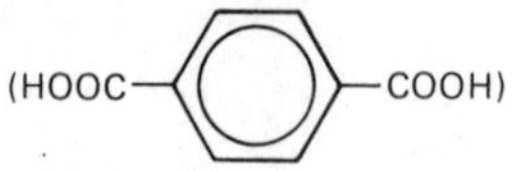

(HOOC—C₆H₄—COOH)

Dacron, Kodel, and Vycron, and these "polyesters" are now the major fibers used worldwide.

11.2 Molding Resins

Thermoplastic polyester resins, like nylons, were developed originally for use as fibers but are now used as engineering resins. The first polyester engineering resin (PET; Rynite) was similar to the crystalline polyester fiber but used sodium compounds as nucleating agents to promote rapid crystallization and a dibenzoic acid ester of neopentyl glycol which lowered the T_g and increased polymer chain mobility.

PET (Celanex, Valox) has a heat deflection temperature under load (DTUL) of 107°C (225°F) and can be used continuously at this temperature. PET engineering plastic can be reinforced with 30 percent fiberglass to produce a plastic with a DTUL value of 227°C (440°F). PET is resistant to dilute nonoxidizing acids, such as hydrochloric acid, aqueous alkalies, aqueous salt solutions, and many solvents, such as ethanol, gasoline, and benzene at moderate temperatures.

The flexibility of thermoplastic polyesters has been enhanced by increasing the number of methylene groups in the repeating units. PBT has a lower DTUL value [66°C (150°F)] than PET, but this value can be increased to 199°C (390°F) by reinforcing with 30 percent fiberglass. The chemical resistances of PBT and PET are similar. Both may be used in long-term service at elevated temperatures.

PET continues to be the major fiber throughout the world, and the major outlet for PET molding resins is for soft drink bottles. Nevertheless, a significant percentage of the 2,600,000 t of PET produced annually in the United States is used as an engineering thermoplastic. In 1988, 914,000 t of PET and PBT were used as engineering thermoplastics.

The demand for PET fiber has stabilized, but there is a need for more PET for bottles and engineering resin applications. The presence of 3 ppm acetaldehyde hinders the use of PET bottles as water containers, but this contaminant is tolerated in soft drink bottles.

Over 100,000 t of fiberglass-reinforced PET was injection-molded in 1987, and it is anticipated that this volume will triple in 1991. Because of its high heat deflection temperature and dimensional stability, PET was selected for the roof rack of the Taurus automobile. Coextruded crystalline and amorphous PET is being used to make containers for nonrefrigerated shelf-stable entrees that can be heated in the container in a microwave oven. An injection-moldable high-barrier PET (Selar) is commercially available.

The rate of crystallization of PET was accelerated in the early 1960s by adding talc. This rate is now accelerated by the use of other nucleating agents, such as low-molecular-weight polypropylene. The molding temperature of PET has been lowered by the use of sodium salts, plasticizers, and lubricants.

In 1988, 60,000 t of PET molding resins was produced in the United States and 66 percent of this volume was used in electrical and electronic applications. There is an increased interest in the more expensive PBT because of its rapid crystallization at 230°F and its increased use in alloys and blends.

11.3 Physical Properties of Engineering Polymers

The typical stress-strain curves for plastics, which were prepared by T. Carswell and H. Nason in 1944, are shown in Fig. 11.1. The properties of PET and PBT may be illustrated by the hard and strong d class. The toughness of these plastics is related to the area under the curve. The toughness may also be quantified by the Izod notched impact test data shown in Table 11.2.

The stress-strain modulus of most unreinforced engineering polymers is in the order of 3×10^5 lb/in^2. A small drop in modulus occurs

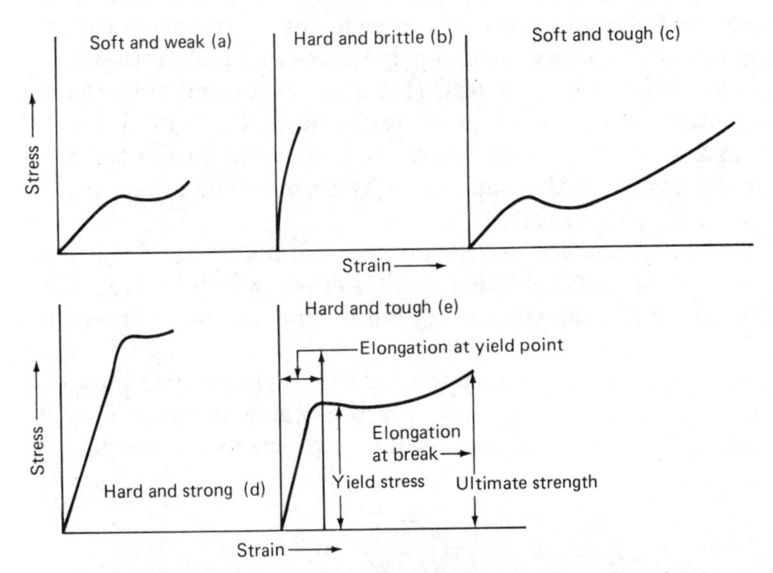

Figure 11.1 Typical stress-strain curves for plastics. (*From* "Introduction to Polymer Chemistry" *by R. Seymour, McGraw-Hill, New York, 1971. Used with permission of McGraw-Hill Book Company.*)

at T_g in crystalline plastics, such as PET and PBT, but the most dramatic drop is at T_m.

In general, a threshold DP value of 50 is required for good structural properties in PET and PBT. Some improvement is noted with higher molecular weights, but the increased energy required for processing highly viscous polymers is seldom justified.

The degree of crystallinity in PET and PBT is related to the efficiency of the nucleating agents present and to the rate of cooling. These crystalline domains are also oriented in the direction of extrusion or injection molding. Of course, the increase in physical properties associated with this alignment of polymer chains is accompanied by a decrease in these properties in the transverse direction.

In addition to the short-term response, which is shown in tables on polymer properties, one should also acquire information on long-term effects of load on physical properties (creep). Creep data may be extrapolated on log-log graphs from a series of data for increasing time periods.

11.4 Effect of Molecular Structure on Properties of Polyesters

In general, the physical and thermal properties of polyesters are related to the relative ratio of ester groups to methylene groups in the polymer chain and to the geometry, polarity, and segmental mobility of the repeating units. The properties are also affected by the intermolecular forces between chains, but these forces in polyesters are less than those in more strongly bonded polyamides and polyurethanes.

Amorphous PET has a T_g of 74°C (165°F), and this value increases as the percentage of crystalline domains increases. The crystalline T_m ranges from 254 to 271°C (490 to 520°F), depending on the thermal history of the PET and the amount of by-product diethylene glycol [$HO(CH_2)_2O(CH_2)_2OH$] present.

The T_m value decreases as the number of methylene groups in polyalkylene terephthalate increases. This effect, which is typical for other engineering plastics containing methylene groups, is shown in Table 11.1.

The T_m values for polyalkylene isophthalates are lower than those for polyalkylene terephthalates, and much of the degree of crystallinity of these polymeric isomers is lost as the value of x increases to 5 or 6.

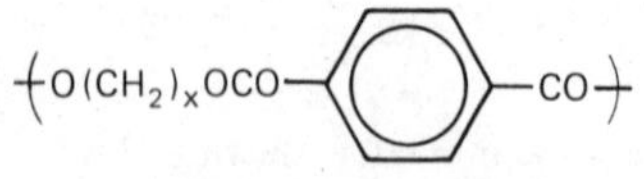

$$\left[O(CH_2)_xOCO-C_6H_4-CO \right]$$

Polyalkylene terephthalate

TABLE 11.1 Effect of Number of Methylene Groups on T_m of Polyalkylene Terephthalates

Number of methylene groups	T_m
2	510
3	450
4	448
5	273
6	309
7	185
8	270
9	203
10	257

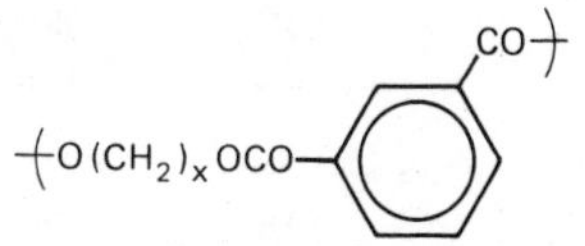

Polyalkylene isophthalate

PET, PBT, and other polyalkylene terephthalates have good dimensional stability, lubricity, and surface gloss. PET and PBT will burn, if ignited, but flame-retardant grades with UL 94 V-O ratings are commercially available. The properties of PET and PBT are shown in Table 11.2.

11.5 Other Thermoplastic Polyesters

In addition to the commercial PET plastics [Petlon (Mobay), Petra (Allied-Signal), Rynite (Du Pont), and Valox (General Electric)], blends and copolymers are commercially available.

Blends of PET and elastomers, such as Valox and VCT (chemically tough), have high impact values (16 ft · lb/in notched Izod). Blends of PET and PBT are less crystalline and more readily moldable. An injection-moldable clear polyester (Kodel) is produced by the condensation of 1,6-dihydroxycyclohexane and terephthalic acid.

Polyarylates, which are completely aromatic, were produced independently by A. Conix, M. Levin and S. Temin, and W. Eareckson in

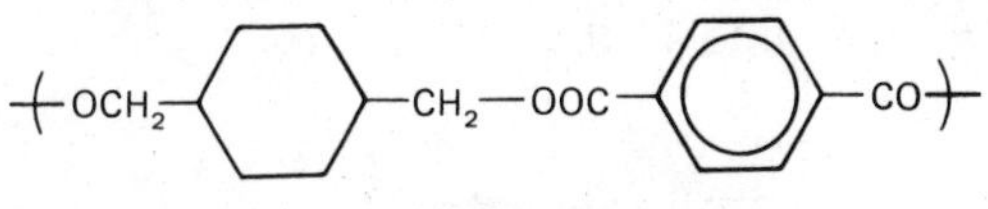

Kodel

the late 1950s. These high-performance plastics are produced by the condensation of BPA with a mixture of isophthalic and terephthalic acids.

TABLE 11.2 Properties of Typical Thermoplastic Aromatic Polyesters

Property, units*	PET	PET + 30% fiberglass	PBT	PBT + 30% fiberglass
Melting point T_m, °F	250	250	245	245
Glass transition temp. T_g, °F	75	—	—	—
Processing temp., °F	550	550	475	475
Molding pressure, 10^3 lb/in^2	5	10	2	10
Mold shrinkage, 10^{-3} in/in	20	5	15	6
Heat deflection temp. under flexural load of 264 lb/in^2, °F	225	440	150	390
Maximum resistance to continuous heat, °F	180	400	130	370
Coefficient of linear expansion, 10^{-6} in/(in · °F)	30	25	35	10
Compressive strength, 10^3 lb/in^2	10	20	11	20
Izod impact strength, ft · lb/in of notch	0.4	1.8	1.0	1.2
Tensile strength, 10^3 lb/in^2	8	16	8	2.1
Flexural strength, 10^3 lb/in^2	16	32	14	25
Percent elongation	150	4	200	3
Tensile modulus, 10^3 lb/in^2	500	1350	350	1400
Flexural modulus, 10^3 lb/in^2	400	1400	370	1000
Rockwell hardness	M96	M100	M74	M90
Specific gravity	1.35	1.6	1.35	1.5
Percent water absorption	0.1	0.07	0.1	0.05
Dielectric constant	3.5	3.5	3.5	3.5
Dielectric strength, V/mil	450	500	450	450
Resistance to chemicals at 75°F:†				
Nonoxidizing acids (20% H_2SO_4)	S	S	S	S
Oxidizing acids (10% HNO_3)	Q	Q	Q	Q
Aqueous salt solutions (NaCl)	S	S	S	S
Polar solvents (C_2H_5OH)	S	S	S	S
Nonpolar solvents (C_6H_6)	S	S	S	S
Water	S	S	S	S
Aqueous alkaline solutions (NaOH)	S	Q	S	Q

*lb/(in^2 · 0.145) = KPa (kilopascals); ft · lb/(in of notch · 0.0187) = cm · N/cm of notch.
†S = satisfactory; Q = questionable; U = unsatisfactory.

Polyarylates are produced by Itika in Japan and marketed under the trade name Ardel by Amoco in the United States. The trade name for the Hoechst Celanese arylate is Durel.

Another commercial aromatic polyester (Ekkcel) is produced by the condensation of *p*-hydroxybenzoic acid. The properties of typical arylates are shown in Table 11.3.

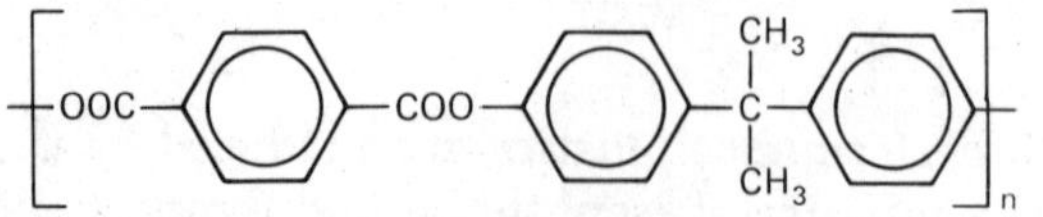

OOC—⟨benzene⟩—COO—⟨benzene⟩—C(CH_3)(CH_3)—⟨benzene⟩—]$_n$

Polyarylate

[—⟨benzene⟩—COO—]$_n$

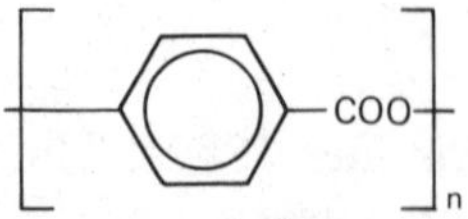

Ekonol

TABLE 11.3 Properties of Typical Arylates

Property, units*	Polyarylates	Polyarylates + 30% fiber-glass	Ekkcel
Melting point T_m, °F	—	—	—
Glass transition temp. T_g, °F	190	190	—
Processing temp., °F	700	700	--
Molding pressure, 10^3 lb/in^2	15	15	--
Mold shrinkage, 10^{-3} in/in	7	1	9
Heat deflection temp. under flexural load of 264 lb/in^2, °F	340	355	345
Maximum resistance to continuous heat, °F	310	325	325
Coefficient of linear expansion, 10^{-6} in/(in · °F)	25	15	28
Compressive strength, 10^3 lb/in^2	12	24	18
Izod impact strength, ft · lb/in of notch	4.5	2.5	1.0
Tensile strength, 10^3 lb/in^2	10	20	14
Flexural strength, 10^3 lb/in^2	12	32	17
Percent elongation	60	2	8
Tensile modulus, 10^3 lb/in^2	30	40	—
Flexural modulus, 10^3 lb/in^2	275	—	700
Rockwell hardness	M90	M95	—
Specific gravity	1.2	1.45	1.4
Percent water absorption	0.5	0.6	0.1
Dielectric constant	3.3	0.6	0.1
Dielectric strength, V/mil	400	400	400
Resistance to chemicals at 75°F:†			
Nonoxidizing acids (20% H_2SO_4)	S	S	S
Oxidizing acids (10% HNO_3)	Q	Q	Q
Aqueous salt solutions (NaCl)	S	S	S
Polar solvents (C_2H_5OH)	S	S	S
Nonpolar solvents (C_6H_6)	S	S	S
Water	S	S	S
Aqueous alkaline solutions (NaOH)	S	Q	S

*lb/(in^2 · 0.145) = KPa (kilopascals); ft · lb/(in of notch · 0.0187) = cm · N/cm of notch.
†S = satisfactory; Q = questionable; U = unsatisfactory.

11.6 Crystallinity

Because of a small degree of flexibility in the polymer chain containing tetramethylene [$(CH_2)_4$], PBT crystallizes at a faster rate than the less flexible PET. PET crystallizes readily at a temperature 46°C (115°F) lower than that at which PET does. However, the polyesters produced by the condensation of terephthalic acid and propylene glycol [$HO(CH_2)_3OH$] and 1,3-butanediol [$HO(CH_2)_2CHOHCH_3$] are amorphous.

The rate of crystallization of PBT, PET, and copolyesters is enhanced by the addition of nucleating agents, such as talc. The melting points of the copolyesters are lowest for 50-50 mixtures of the reactants [179°C (355°F)] and increase linearly as the concentration of either reactant is increased.

The rate of crystallization of PET is related to the quench temperature, and the rate decreases below the critical temperature T_K, at which the rate of crystallization is retarded because of poor molecular chain mobility. The T_K of PET is 180°C (356°F) and is related to T_g and T_m as follows:

$$T_K = \frac{T_g + T_m}{2}$$

Note: These temperatures are in kelvin (K).

PET is readily molded at temperatures below T_g, but these molded parts must be used at temperatures below T_g. PET is a very viscous melt which is difficult to mold above T_g. This problem was solved by the use of a two-part crystallization system in which the dibenzoate of neopentyl glycol was used as a chemical nucleating agent instead of talc or glass flakes, which are physical nucleating agents. Sodium or potassium salts of polymeric acids, such as ionomers (ethylene–methacrylic acid copolymers), are also added to increase the molecular chain mobility and allow crystallization of the modified PET at a lower mold temperature, and fully crystallized molded parts can be produced in this manner.

When stabilized by the addition of UV stabilizers, PET and PBT have excellent resistance when exposed to sunlight. Because they undergo a photo-Fries rearrangement, which produces *o*-hydroxybenzophenone groups which are UV stabilizers, arylates are extremely resistant to the effects of UV radiation. However, the clear films become yellow after long-term exposure to sunlight.

11.7 Reinforcements

While the properties of polyesters with 30 percent fiberglass are shown in Tables 11.2 and 11.3, commercial grades with fiberglass reinforcements in the range of 15 to 55 percent are available commercially. Injection-molded PET or PBT parts tend to be anisotropic, i.e., the polymer chains are aligned in the direction of flow, and this effect is enhanced when fiberglass is present in the composite. This alignment causes nonuniform shrinkage, which results in curling of thin sections. This tendency may be reduced by the addition of mica, which is a platelike filler.

The toughness, tensile strength, and notched impact resistance are increased by fiberglass reinforcements. Several commercial flame-retardant fiberglass-reinforced aromatic polyesters have UL 94 V-0 ratings.

11.8 Thermal and Flame Resistance

The thermoplastic aromatic polyesters have exceptionally good thermal resistance for thermoplastics. As shown by data in Table 11.4, Ardel D-100 polyarylate, which has a UL temperature index of 130°C (266°F), retains many of its mechanical properties after long-term aging at 149°C (300°F).

While the values shown in Table 11.5 should not be used to predict performance under actual fire conditions, flame-retardant unfilled aromatic polyesters, such as Ardel D-100, have shown excellent resistance to burning in combustion tests.

11.9 Electrical Properties

As shown by the data for Ardel D-100 in Table 11.6, thermoplastic aromatic polyesters have excellent electrical properties.

11.10 Processing of Polyesters

Polyesters are hygroscopic, and if not removed, any absorbed moisture in the molding powder may result in hydrolytic degradation in the molding or extrusion process. Hence, it is essential that the molding powder be dried at 135°C (275°F) to a moisture content of less than

TABLE 11.4 Effects of Temperature on Mechanical Properties of Ardel Polyarylate D-100

	Hours aged at 300°F (tested at 72°F)		
Property	0	500	2500
Tensile modulus, lb/in^2	290,000	275,000	310,000
Tensile strength, lb/in^2	9,500	10,700	11,400
Tensile elongation at rupture, %	50	12	10.5
Tensile elongation at yield, %	8	7.6	7.5
Flexural strength, lb/in^2	11,000	—	—
Flexural modulus, lb/in^2	310,000	—	—
Izod impact strength (1/8 in), ft · lb/in of notch	4.2	3.6	2.0
Falling dart impact (1/8-in plaque), ft · lb	> 68	> 68	> 68
Tensile impact, ft · lb/in	130	117	182

TABLE 11.5 Combustion Data for Ardel D-100

Property	Test method	Value
Flammability, 1/16-in	UL 94	V-0
Limiting oxygen index, 1/4-in	D 2863	36
Autoignition temp.	D 1929	545°C
Flame spread index, 1/16-in	E 162	1.87
Heat release, 2.5 W/cm^2, 1/16-in	OSU	
Max rate of release		4 kcal/s · m^2
Total heat release		142 kcal/m^2
Average burning time, 1/8-in	D 635	2.6 s
Average extent of burning, 1/8-in		0.75 in
Smoke density, flaming mode vertical sample, 1/16-in	NBS	
D_s at 1.5 min		2.3
D_s at 4 min		34.0
D_s max		109.0
Time to D_s max		10.9 min
Time to D_s = 16		3.1 min

TABLE 11.6 Electrical Properties of Ardel D-100

Property	ASTM method	At 72°F, 50% RH*
Dielectric constant	D 150	3.34
At 60 Hz		3.34
At 1000 Hz		3.32
At 1,000,000 Hz		3.30
Dissipation factor	D 150	
At 60 Hz		0.002
At 1000 Hz		0.004
At 1,000,000 Hz		0.02
Volume resistivity	D 257	3×10^{16} Ω · cm
Surface resistivity, 5-mil film	D 257	2×10^{17} Ω
Dielectric strength, step by step	D 149	400 V/mil
Arc resistance	D 495	125 s

*RH = relative humidity.

0.02 percent and that dehumidifier drier equipment be used on the hopper.

While the processing conditions shown in Tables 11.2 and 11.3 may be used as general guidelines, injection cylinder temperatures, processing conditions, and mold dimensions provided by the resin supplier should be followed.

11.11 Applications

Thermoplastic aromatic polyesters are used for electrical and electronic applications, such as TV components, ignition rotors, and electric motor end caps, and for nonelectrical automotive applications, such as air deflectors, grills, fenders, and wheel covers. These plastics are also used for iron handles, hair driers, hair drier housings, lawn mowers, lawn mower housings, surgical instrument handles, and microwave cookware.

PET film, obtained by melt extrusion (Mylar, Melinex, Celanar), is widely used for photographic film and packaging. Blow-molded PET is the most widely used plastic for bottles of all types.

11.12 Glossary

acetaldehyde H_3CCHO

adipic acid $HOOC(CH_2)_4COOH$

aliphatic A derivative of a petroleum-derived hydrocarbon.

alkylene A general term for —R— groups.

amorphous Noncrystalline.

anisotropic Having nonuniform properties in different areas of a plastic specimen.

Ardel A polyarylate.

aromatic A derivative of benzene.

bisphenol A $HOC_6H_4C(CH_3)_2C_6H_4OH$

1,3-butane diol $HO(CH_2)_2CHOHCH_3$

Carothers The inventor of nylon and neoprene.

Carothers' equation $\overline{DP} = \frac{1}{1 - p}$

Celanar A polyester film.

Celanex A polyester.

copolyester One with two different polyester groups in the repeating units.

creep The time-dependent strain exhibited by a plastic after long-time stress.

Dacron A polyester.

1,6-Dihydroxycyclohexane HO—⬡—OH, ⬡ = cyclohexane

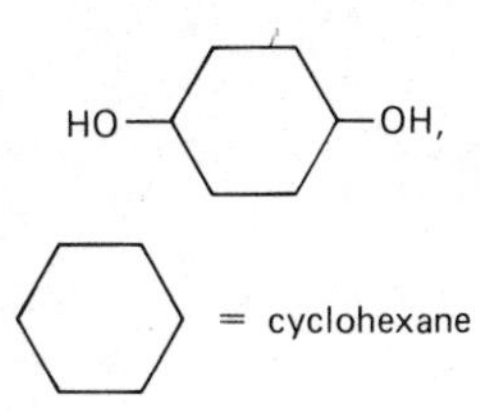

dicarboxylic acid HOOCRCOOH

diethylene glycol $HO(CH_2)_2O(CH_2)_2OH$

$\overline{DP}$ The average degree of polymerization, i.e., the number of repeating units in a polymer.

DTUL The heat deflection temperature under load.

Durel An arylate.

Ekkcel An arylate.

Ekonal An arylate.

engineering resin A resin that has high tensile strength and can be used in place of metals.

ester group —COOR

flame retardant An additive which retards the rate of burning of a plastic.

force, intermolecular An attractive force between polymer chains.

glass transition temperature The temperature at which a brittle polymer softens and becomes flexible when heated.

Glyptal A polyester based on the condensation of glycerol and phthalic anhydride.

hydrolytic degradation A loss in molecular weight resulting from hydrolysis of ester groups in the polymer chain.

hygroscopic Moisture-absorbing.

ionomer A copolymer of ethylene and methacrylic acid.

isophthalic acid Meta-phthalic acid.

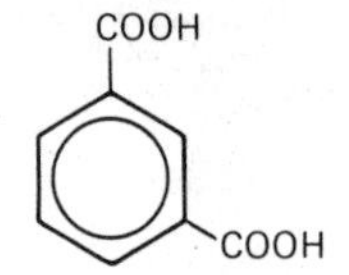

isotropic Having properties independent of the direction of the test.

Izod test An impact test in which a falling pendulum strikes a cantilevered specimen, which is usually notched.

kelvin The absolute temperature scale; 273 K = 0°C = 32°F.

Kienle The inventor of alkyd polyesters.

Kodel A polyester.

limiting oxygen index The highest percentage of oxygen in an oxygen-hydrogen mixture in which a specimen will not burn in a candle-type flammability test.

LOI Limiting oxygen index.

lubricity Slipperiness.

Melinex A polyester film.

methacrylic acid $H_2C{=}C(CH_3)COOH$

methylene group CH_2

mica A platelike filler.

modulus of elasticity The stress/strain ratio.

Mylar A polyester film.

neopentyl glycol $HOCH_2C(CH_3)_2CH_2OH$

nucleating agent One that serves as a site for initiating the formation of crystals.

p The extent of polymerization or yield.

PBT Polybutylene terephthalate.

PET Polyethylene terephthalate (polyester).

Petlon A polyester.

Petra A polyester.

photo-Fries rearrangement A UV-initiated rearrangement of arylates which produces hydroxybenzophenone groups.

***p*-Hydroxybenzoic acid** HO—⌬—COOH

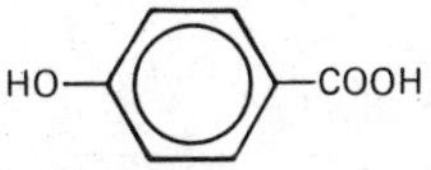

plasticizer A flexibilizing additive.

polyarylates Completely aromatic polyesters.

polybutylene terephthalate $\left[O(CH_2)_4—OOC—⌬—CO\right]$

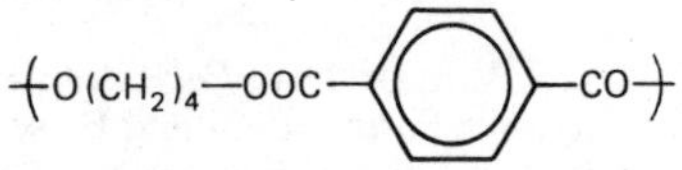

polyester $\left(OROOCRCO\right)$

polypropylene $\left[CH(CH_3)CH_2\right]$

ppm Parts per million.

propylene glycol $HO(CH_2)_3OH$

Rynite A polyester.

Selar A high-barrier polyester.

strain A change in length, i.e., elongation.

stress A force acting on a solid material, usually tensile or pulling force.

talc An acicular (needlelike) filler which serves as a nucleating agent in PET.

terephthalic acid HOOC—⌬—COOH

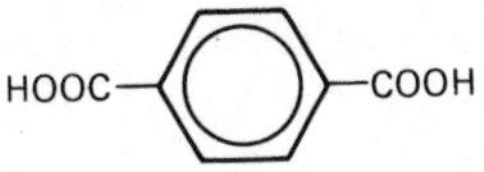

Terylene A polyester.

T_g Glass transition temperature.

thermoplastic A plastic that can be reversibly softened by heat and then cooled to produce a hard solid.

thermoset A plastic that cannot be softened by heat without decomposition.

T_K The critical temperature.

T_m The melting point or first-order transition.

UL Underwriters' Laboratory.

UL 94 V-0 No afterglow after the removal of a flame at the bottom of a vertical test specimen after two 10-s intervals.

UV Ultraviolet.

Valox A polyester.

VCT Valox, chemically tough.

Vycron A polyester.

11.13 References

Deyrup, E. J., D. W. Fox, D. McNally, and J. S. Gall: Chapters 8, 9, 10 in *High Performance Polymers: Their Origins and Development,* Elsevier Science Publishing Co., Inc., New York, 1986.

Goodman, I.: "Polyesters," in J. I. Kroschwitz (ed.): *Encyclopedia of Polymer Science and Engineering,* vol. 12, Interscience Publishers, John Wiley & Sons, Inc. New York, 1988.

Goodman, I., and J. A. Rys: *Saturated Polyesters,* Plastics and Rubber Institute, London, 1965.

Korshak, V. V., and S. Vinogradova: *Polyesters,* Pergamon Press, Oxford, 1965.

Lawrence, J. R.: *Polyester Resins,* Reinhold Book Corporation, New York, 1968.

Margolis, J. M., and J. R. Irish: Chapters 1, 2 in *Engineering Thermoplastics,* Marcel Dekker, Inc., New York, 1985.

Moncrieff, R. W.: Chapter 24 in *Man Made Fibers,* Newnes-Butterworths, London, 1975.

Seymour, R. B., and C. E. Carraher: *Structure-Property Relationships in Polymers,* Plenum Press, Plenum Publishing Corporation, New York, 1984.

Seymour, R. B., and C. E. Carraher: *Polymer Chemistry: An Introduction,* Marcel Dekker, Inc., New York, 1988.

Chapter

12

Aliphatic and Aromatic Polyamides (Nylons)

12.1 Introduction

Since nylon was available for parachutes in the late 1930s and as hosiery in the late 1940s, most readers cannot remember a world without this important polymer. Proteins have been present since the beginning of life on earth, and synthetic PAs were synthesized by Gabriel and Nobel laureate Emil Fischer in the early 1900s. However, nylon, which was the world's first truly synthetic fiber, was unknown until W. C. Carothers synthesized polyhexamethyleneadipamide, nylon 6,6, $+NH(CH_2)_6NHCO(CH_2)_4CO+$, in the early 1930s.

Linear polyester fibers, which were described in Chap. 11, were also produced by Carothers, but he shelved the investigation of these low-melting-point aliphatic polyesters in order to devote his research efforts to the higher-melting-point PAs. All his diadic PAs were prepared by the thermal dehydration of the salt of an aliphatic diamine (H_2NRNH_2) and an aliphatic dicarboxylic acid (HOOCRCOOH). The numbers following the generic nylon name indicate the number of carbon atoms in the diamine and dicarboxylic acid in that order. Thus, the polymer obtained by heating the salt formed by the reaction of hexamethylenediamine [$H_2N(CH_2)_6NH_2$] and adipic acid [$HOOC(CH_2)_4COOH$] is called nylon 6,6.

Other diadic PAs, such as nylon 6,10 and nylon 12,12, are commercially available, but nylon 6,6 continues to be the major PA fiber and engineering polymer. In 1988, 254,000 t of nylon 6,6 engineering polymers was used in the United States. The principal American producers of nylon 6,6 are Du Pont and Monsanto.

Monadic nylons, such as nylon 6, $+NH(CH_2)_5CO+$, were produced by P. Schlack in 1937 by the polymerization of caprolactam:

$(\overline{CH_2(CH_2)_4NHCO})$

This synthetic PA, which was commercialized by IG Farbenindustrie and sold under the trade name Perlon L, is now called by the generic name nylon 6.

Nylon 6 was produced sporadically in Germany during World War II, but the production facilities in East Germany were dismantled and shipped to the Soviet Union after the armistice. Nylon 6, which is used both as a fiber and as an engineering polymer, is now produced by Hoechst in Germany, by Allied Chemical (Caprolam), by Toyo (Amilon) in Japan, and in the Soviet Union (Kapron). Nylon 11, $+NH(CH_2)_{10}CO+$, and nylon 12, $+NH(CH_2)_{11}CO+$, are also commercially available.

While nylon 6 and nylon 6,6 are ideal fibers for many applications, they soften when heated above T_m. This deficiency for high-temperature applications was overcome in 1967 by J. Hill, S. Kwolek, and P. Morgan of Du Pont, who produced and patented aromatic nylons (nylon 6-T). This high-melting-point polymer [T_m = 404°C (760°F)] was produced by the condensation of hexamethylenediamine and terephthalic acid. An aromatic nylon (Nomex), which is now called by the generic name aramid, was produced by the condensation of *meta*-phenylenediamine ($H_2NC_6H_4NH_2$) and isophthaloyl chloride ($ClCOC_6H_4COCl$). Other aramids, such as poly-*p*-benzamide [$+C_6H_4—CONH+$] are commercially available under the trade name Kevlar.

12.2 Processing of Nylon 6,6

Nylon 6,6 is the world's most widely used engineering polymer. This plastic is a member of the synthetic polyamide family, which includes other aliphatic and aromatic nylons. It is important to note that PAs are unrelated to the thermoset amino plastics (ureas and melamines) described in Chap. 2. The name "nylon" is a generic term originally coined to stand for "no run" hosiery.

Since nylon 6,6 is a crystalline polymer, its degree of crystallinity must be controlled by quenching via cooled molds [71 to 91°C (160 to 195°F)]. The recommended gate and runner sizes are related to the thickness of the molded part. Thus, a gate with a diameter as large as 0.2 in and a long (21 in) hot runner with a large diameter of 0.4 in are used for thick parts (0.5 in), and the diameters are much smaller for thinner parts, i.e., with diameters less than 0.2 in. The recommended molding temperature is in the 260 to 279°C (500 to 535°F) range, which borders on the T_m.

Slightly higher temperatures are used for the extrusion of nylon 6,6. A temperature of 290°C ± 3°C (555°F ± 5°F) is preferred, and high-molecular-weight, i.e., high melting viscosity, nylon 6,6 is usually specified for the extrusion of tubing. It is essential that the extrudate be pulled rapidly from the die using a drawdown ratio of about 2.5:1. The rate of pipe extrusion is slower than that of tubing extrusion, and a temperature of about 20°C (36°F) above T_m should be maintained in the extruder.

The physical and mechanical properties of nylon 6,6 are related to the degree of crystallinity. Hence, in all processing techniques, nucleating agents, such as fumed silica, PET, or lithium fluoride (LiF) (0.5%) should be present to promote crystallization with its inherent increase in tensile strength, flexural strength, and modulus, and corresponding decrease in impact resistance and elongation.

Nylon 6,6 is extruded as film (8%), sheet rod and tube (6%), wire and cable (5.5%), and filaments (4%). Molded nylon 6,6 is used for electrical and electronic applications (12%), consumer products (10%), industrial uses (10%), transportation (8.5%), and appliances (3%).

12.3 Properties of Nylon 6,6 and Nylon 6,12

The various commercial nylons differ in melting points and moisture absorption, but their chemical resistance is a function of the resistance of the amide groups ($CONH_2$) which are present in the polymer chain. Nylon 6,6 is an abrasion-resistant hygroscopic, high-melting-point, crystalline polymer. Its T_m and water absorption are 266°C (510°F) and 1.5 percent, respectively. Nylon is marketed under the trade names Elvamide, Zytel, and Rilsan.

Amorphous nylon, which is produced by the condensation of several reactant homologues or by quenching, is a clear polymer that is less heat resistant than nylon 6,6. Nylon 6,6 may be reinforced by fiberglass or inorganic fillers or toughened by the addition of elastomers. The toughened nylon has outstanding resistance to impact and to stress cracking. In an arbitrary empirical rating, MacDermott ranked super tough nylon above all the other engineering plastics.

Nylon is resistant to aqueous salts, alkalies, and dilute solutions of nonoxidizing acids. It is attacked by concentrated mineral acids, such as hydrochloric acid, by oxidizing acids, such as nitric acid, by hydrogen peroxide, and by chlorine bleaching solutions. Nylon swells in concentrated formic and acetic acid but is resistant to other organic acids.

Du Pont and Monsanto produce mineral-filled and fiberglass-reinforced nylon 6,6 under the trade names Zytel and Vydyne, respectively. Long-fiber-reinforced nylon 6,6 (Verton), which has a much higher impact resistance than short-fiber-filled nylon 6,6, is being produced by pultrusion techniques.

The T_m and percent water absorption decrease as the number of methylene groups in the repeating units of nylon is increased. The moisture-absorption values of nylon 6,6, nylon 6,9, and nylon 6, 12 at 100 percent humidity are 8, 4, and 3 percent, respectively. The modulus of nylons is decreased under conditions of high humidity. The absorbed moisture has an adverse effect on the dimensional stability but does enhance the flexibility of molded or extruded parts.

The properties of nylon 6,6 are tabulated in Table 12.1.

12.4 Properties of Nylon 6, Nylon 7, Nylon 11, and Nylon 12

Nylon 6, which was produced by P. Schlack in 1937, is made by the anionic polymerization of caprolactam. It is being produced in the United States by Allied-Signal. Motto, Schnell, and Wichterle discovered the advantage of adding acylating agents to caprolactam in 1975, and this led to the commercial "monomer casting" of large parts, such as bumpers and fuel tanks, by Monsanto, Bayer, and BASF. BASF estimates the worldwide production of nylon 6 to be about 300,000 t. Nylon 7 and nylon 12 were produced by Zeltner in 1938. These resins are less moisture sensitive than nylon 6,6 and nylon 6. Because of the lower T_m values, all these monadic nylons must be molded and extruded at lower temperatures than nylon 6,6. The recommended injection-molding and extrusion temperatures for nylon 6 are about 282°C (540°F), respectively. The extrusion temperature for nylon 12 is about 210°C (410°F). The procedures outlined for nylon 6,6 should be used as a guide for processing other nylons, but it is important to follow the suppliers' directions.

Nylon 6 is readily rotomolded and blow-molded. Hollow parts are produced by rotating a heated mold containing a measured quantity of nylon 6 simultaneously in two directions perpendicular to each other. After a desired thickness of the molded part is obtained from the molten polymer, the mold is cooled and unloaded. In the blow-molding process, a short tube (parison) is heated and one end is closed before forcing the heated polymer against the mold surface by means of a compressed gas in a closed split mold. The finished part is removed after it is cool enough to maintain its shape.

The monomer casting process (Nylacast) has been modified so that the anhydrous molten caprolactam with an initiating system is molded at 141°C (285°F) and cooled. A mixture of a methyllactam and a polyacetyllactam is used as the initiation system for the in situ polymerization in the RIM process. Powdered nylon 6, nylon 11, and

TABLE 12.1 Properties of Typical Diadic Nylons

Property, units*	Nylon 6,6	Nylon 6,12	Nylon 6,6 + 30% long fiberglass	High-impact nylon 6,6 (rubber-modified)
Melting point T_m, °F	500	400	500	490
Glass transition temp. T_g, °F	—	—	—	—
Processing temp., °F	550	500	550	550
Molding pressure, 10^3 lb/in^2	20	10	15	15
Mold shrinkage, 10^{-3} in/in	10	10	3	15
Heat deflection temp. under flexural load of 264 lb/in^2, °F	180	160	500	160
Maximum resistance to continuous heat, °F	150	135	450	135
Coefficient of linear expansion, 10^{-6} in/(in · °F)	40	35	15	50
Compressive strength, 10^3 lb/in^2	14	10	35	—
Izod impact strength, ft · lb/in of notch	1.0	2.0	4.0	6.0
Tensile strength, 10^3 lb/in^2	13	7.5	28	7.5
Flexural strength, 10^3 lb/in^2	16	10	35	30
Percent elongation	100	200	3.0	200
Tensile modulus, 10^3 lb/in^2	400	250	1400	800
Flexural modulus, 10^3 lb/in^2	400	300	1400	1100
Rockwell hardness	M83	M75	E60	R115
Specific gravity	1.14	1.08	1.4	1.1
Percent water absorption	1.2	1.0	1.0	1.0
Dielectric constant	4.0	4.0	4.0	4.0
Dielectric strength, V/mil	600	400	500	400
Resistance to chemicals at 75°F:†				
Nonoxidizing acids (20% H_2SO_4)	U	U	U	U
Oxidizing acids (10% HNO_3)	U	U	U	U
Aqueous salt solutions (NaCl)	S	S	S	S
Polar solvents (C_2H_5OH)	Q	Q	Q	Q
Nonpolar solvents (C_6H_6)	S	S	S	S
Water	S	S	S	S
Aqueous alkaline solutions (NaOH)	S	S	S	S

*lb/(in^2 · 0.145) = KPa (kilopascals); ft · lb/(in of notch · 0.0187) = cm · N/cm of notch.
†S = satisfactory; Q = questionable; U = unsatisfactory.

nylon 12 may be used for powder coating of metal parts. Coatings may be applied by dipping a hot metal part in a fluidized bed of finely divided nylon or electrostatic spraying of the powder. The semifinished coated parts are then heated to ensure a continuous film.

The dimensional stability of nylons increases as the number of methylene groups in the repeating units increases. All PAs undergo autooxidation at elevated temperatures. The rate of this degradation

TABLE 12.2 Properties of Typical Monadic Nylons

Property, units*	Nylon 6	Nylon 6 + 30% long fiberglass	Nylon 11	Nylon 12
Melting point T_m, °F	420	420	375	360
Glass transition temp. T_g, °F	—	—	—	—
Processing temp., °F	500	500	450	450
Molding pressure, 10^3 lb/in^2	15	15	10	10
Mold shrinkage, 10^{-3} in/in	10	13	12	12
Heat deflection temp. under flexural load of 264 lb/in^2, °F	165	420	130	120
Maximum resistance to continuous heat, °F	150	380	120	110
Coefficient of linear expansion, 10^{-6} in/(in · °F)	45	14	60	60
Compressive strength, 10^3 lb/in^2	14	24	12	12
Izod impact strength, ft · lb/in of notch	1.0	4.0	2.0	2.2
Tensile strength, 10^3 lb/in^2	12	26	8	7
Flexural strength, 10^3 lb/in^2	15	40	20	20
Percent elongation	200	2	300	325
Tensile modulus, 10^3 lb/in^2	350	1000	185	150
Flexural modulus, 10^3 lb/in^2	350	1200	150	120
Rockwell hardness	R119	M94	R108	R85
Specific gravity	1.13	1.4	1.05	1.02
Percent water absorption	2.0	1.3	0.3	0.25
Dielectric constant	4.0	4.0	4.0	4.0
Dielectric strength, V/mil	400	400	425	450
Resistance to chemicals at 75°F:†				
Nonoxidizing acids (20% H_2SO_4)	U	U	U	U
Oxidizing acids (10% HNO_3)	U	U	U	U
Aqueous salt solutions (NaCl)	S	S	S	S
Polar solvents (C_2H_5OH)	Q	Q	S	S
Nonpolar solvents (C_6H_6)	S	S	S	S
Water	S	S	S	S
Aqueous alkaline solutions (NaOH)	S	S	S	S

*lb/(in^2 · 0.145) = KPa (kilopascals); ft · lb/(in of notch · 0.0187) = cm · N/cm of notch.
†S = satisfactory; Q = questionable; U = unsatisfactory.

can be decreased by the addition of copper salts or hindered amine light stabilizers (HALS). The properties of monadic nylons are tabulated in Table 12.2.

12.5 Effect of Humidity on Nylon

Because of the presence of hydrophilic amide groups, all nylons are affected by moisture, but these effects decrease as one goes from nylon 6 to nylon 12, or nylon 6,6 to nylon 6,12. Thus, while nylon 6 and nylon 6,6 undergo a dimensional change of about 0.65 percent at 50 percent relative humidity, nylon 11 and nylon 12 undergo a dimensional change of about 0.10 percent under these conditions.

The volume resistivity decreases and the dielectric constant increases as the percent humidity increases, and these effects are greatest for nylon 6 and nylon 6,6 and least for nylon 11 and nylon 12. The dielectric constant at 50 percent humidity is as follows: nylon 6 (10), nylon 6,6 (7), nylon 11 (5), and nylon 12 (3).

12.6 Nylon Powder Coating

Articles may be readily coated by nylon powder in a fluidized bed or by electrostatic projection techniques. The latter method, in which nylon powder is passed through a nozzle carrying a high electric potential, is preferred for thin coatings. Because of its lower T_m, which facilitates the aftercoating fusion process, nylon 11 is preferred unless the coating is to be used at relatively high temperatures. Most nylon coatings are approved by the U.S. Department of Agriculture (USDA).

With nylon and other polymer coatings, sharp corners on the coated object should be avoided, i.e., a radius of at least 0.03 in should be present on exterior and interior corners. The space between wires on coated wire mesh must be at least three diameters for lower-gauge meshes. The diameter of holes must be greater than their depth. Small parts should be preheated and dipped into a fluidized bed in a vibrating bowl.

12.7 Aramids

The amide groups in aromatic PAs (aramids) are attached to cyclic rings at both ends of the polymer chain. In general, the para isomer, such as Kevlar (poly *p*-phenylene-diamine-co-terephthalic acid) are rodlike and yield melts with moderately low viscosities. Poly-*p*-benzamide (PBA), which was introduced by Du Pont in 1970, is used as tire cord. Nomex, which was introduced by Du Pont in 1967, is a meta isomer, i.e., poly-*m*-phenyleneisophthalamide. The meta-oriented PAs are chain-folding polymers.

Aramids are characterized by high tensile strength, high modulus, and low elongation. These fibers are difficult to burn and have excellent electrical properties which are maintained at elevated temperatures. These fibers under the trade names Conex, Kevlar, Nomex, and Twaron are used for V-belts, bulletproof vests, filament-wound structures, and membranes and as replacements for asbestos. Aramids are available as paperlike film but are not used as molded resins at the present time.

12.8 Glossary

acid, adipic $HOOC(CH_2)_4COOH$

acid, dicarboxylic HOOCRCOOH

acid, terephthalic $HOOC-C_6H_4-COOH$

acylating agent RCOCl

aliphatic A noncyclic hydrocarbon compound related to methane (CH_4) and similar compounds found in natural gas and petroleum.

amide group —NHCO—

amilon Nylon 6.

amino plastics Thermoset urea and melamine plastics.

amorphous Noncrystalline.

aramid An aromatic polyamide.

aromatic An unsaturated hydrocarbon compound related to benzene.

blow molding The expansion of a heated tube (parison) by air pressure against the inner surface of a mold.

caprolactam Nylon 6.

Carothers, W. C. The inventor of nylon and neoprene.

casting, monomer The polymerization of a monomer, such as caprolactam, in a mold cavity.

diadic Produced by the condensation of a diol and a dicarboxylic acid.

diamine H_2NRNH_2

elastomer Rubber.

Elvamide A nylon.

extrusion A process in which a molten plastic is pushed through an orifice by a rotating screw.

filament A continuous small-diameter extrudate.

Fisher, E. A pioneer protein chemist and Nobel laureate.

fluidized bed A mixture of finely divided polymer and a gas in which hot parts are dipped and coated with the resinous powder.

gate An orifice through which the molten plastic enters the mold in an injection-molding press.

HALS Hindered amine light stabilizers.

hexamethylenediamine $H_2N(CH_2)_6NH_2$

hydrolytic degradation A loss in molecular weight resulting from hydrolysis of ester groups in the polymer chain.

hydrophilic Water-loving.

hygroscopic Water-absorbing.

Kapron A nylon 6.

Kevlar An aramid.

lithium fluoride LiF

methylene group CH_2

monadic Nylon produced from a lactam, i.e., one repeating unit only in the polymer.

Nomex An aramid.

nucleating agent An agent that serves as a site for initiating the formation of crystals.

Nylacast A process for monomer casting of nylon 6.

nylon A synthetic polyamide.

nylon 6 $-[NH(CH_2)_5CO]-$

nylon 6-T An aramid.

nylon 7 $-[NH(CH_2)_6CO]-$

nylon 11 $-[NH(CH_2)_{10}CO]-$

nylon 12 $-[NH(CH_2)_{11}CO]-$

nylon 6,6 $-[NH(CH_2)_6NHCO(CH_2)_4CO]-$

nylon 6,10 $-[NH(CH_2)_6NHCO(CH_2)_8CO]-$

PA Polyamide.

parison A plastic tube closed at one end.

Perlon L A nylon 6.

plasticizer A flexibilizing additive.

polyamide $-(NHCOR)-$

polypropylene $-[CH(CH_3)CH_2]-$

pultrusion The process of pulling a bundle of resin-impregnated filaments through a die.

reaction injection molding A process in which polymerization takes place in the mold cavity.

Rilsan Nylon 11.

RIM Reaction injection molding.

rotomolding The molding of a polymeric powder in a closed rotating mold cavity.

runner A channel through which molten plastic flows before entering the mold in an injection-molding press.

Schlack, P. The inventor of nylon 6.

silica, fumed Finely divided silica produced by the reaction of hydrogen (H_2) on silicon tetrachloride ($SiCl_4$).

strength, flexural Resistance to a bending force.

strength, tensile Resistance to a pulling force.

terephthalic acid 

Terylene A polyester.

thermoplastic A plastic that can be reversibly softened by heat and then cooled to produce a hard solid.

thermoset A plastic that cannot be softened by heat without decomposition.

T_m The melting point or first-order transition.

Verlon A long fiber-reinforced nylon.

Vycron A polyester.

Vydyne A nylon 6,6.

Zytel A nylon 6,6.

12.9 References

Black, W. B., and J. Preston (eds.): *High Modulus Wholly Aromatic Fibers,* Marcel Dekker, Inc., New York, 1973.

Floyd, D. E.: *Polyamide Resins,* Reinhold Publishing Company, New York, 1958.

Kohan, M. I. (ed.): *Nylon Plastics,* John Wiley & Sons, Inc. New York, 1973.

Kohan, M. I., P. Matthias, W. F. Seydl, G. B. Apgar, and M. J. Koskoski: in R. B. Seymour and G. S. Kirshenbaum (eds.): *High Performance Polymers: Their Origin and Development,* Elsevier Science Publishing, New York, 1986.

Lewin, M. (ed): *Handbook of Fiber Science and Technology,* vol. 3, Marcel Dekker, Inc., New York, 1985.

MacDermott, C. P.: *Selecting Thermoplastics for Engineering Applications,* Marcel Dekker, Inc., New York, 1984.

Moncrieff, R. W.: *Man-Made Fibers,* Newnes-Butterworth, London, 1975.

Preston, J., and R. J. Welgos: *Encyclopedia of Polymer Science and Engineering,* vol. 2, Interscience Publishers, John Wiley & Sons, Inc. New York, 1988.

Seymour, R. B. (ed.): *History of Polymer Science and Technology,* Marcel Dekker, Inc., New York, 1982.

Chapter

13

Isocyanate Polymers

13.1 Introduction

Polyurethanes (PURs), like PAs are extremely versatile molecules. The urethane reaction, i.e., the reaction of an alcohol (ROH) and an aryl isocyanate (ArNCO), was described by C. Wurtz in 1849. Prior to the introduction of infrared spectroscopy, chemists depended on this Wurtz reaction to characterize alcohols.

Coincidental with the development of nylon 6,6 by W. C. Carothers, and nylon 6 by P. Schatz, O. Bayer investigated the production of PURs by the reaction of diols (HOROH) and diisocyanates (OCNArCNO). As shown by the following equations, these diisocyanates also react with diamines (H_2NRNH_2) to produce polyureas.

$$n\text{OCNRNCO} + n\text{HOROH} \rightarrow \underbrace{\text{OCN(RNHCOOR)}_n\text{OH}}_{\text{PUR}}$$

$$n\text{OCNRNCO} + n\text{H}_2\text{NRNH}_2 \rightarrow \underbrace{\text{OCN(RNHCONHR)}_n\text{NH}_2}_{\text{polyurea}}$$

The —NHCOO— group is called the urethane group, and the —NHCONH— group is called the urea group.

The end uses for these versatile polymers are as rigid and flexible foam for furniture and mattresses (32%), in automotive applications (20%), in construction (15%), as foam for refrigeration and insulation (6%), as footwear (4%), as packaging (2%), and as coatings, adhesives, and encapsulants (4%). Over 1,300,000 t of polyurethanes was used in the United States in 1988. RIM and cast elastomers accounted for 72,000 and 47,000 t of this production, respectively.

13.2 RIM

Large parts are readily produced in relatively inexpensive molds at relatively low pressures by RIM. The reactants usually consist of a specific quantity of a hydroxyl-terminated polyether or polyester resin and a specific quantity of an aromatic diisocyanate, such as diphenylmethane diisocyanate (MDI). This RIM process is sometimes referred to as liquid injection molding (LIM).

The gates used to fill the mold cavity in RIM allow the transport of a high-velocity stream of viscous reactants into a wide thin film which flows over a dam and is directed to the lowest surface of the mold cavity. A flow viscosity of less than 5 ft/s and a film thickness of less than 0.1 in are recommended. The mold cavity must be designed to permit venting of entrapped air and to control the temperature [±2.8°C (5°F)]. The pressure is in the order of 150 lb/in^2, and the reaction time is usually less than 3 s. Parts weighing up to 10 lb can be removed after a residence time of less than 2 min.

The finished parts, which are usually microcellular, with a specific gravity of 0.7 to 0.8, can be strengthened by the incorporation of reinforcing agents, such as milled glass fibers.

Thermoset PUR moldings may be produced by using a small amount of a triol along with the diol in the reactants. The ether-based PURs are more hydrolytically stable than those based on esters. However, antihydrolytic agents, such as polycarbodiimide $[-(RN{=}C{=}N)-]$, may be added. Flame retardants, such as trichloroethyl phosphate, $O{=}P[O(CH_2)_2Cl]_3$, and stabilizers, may also be present.

13.3 Other Reactants

Polyols may be obtained from BASF, Dow, Mobay, Olin, Texaco, or Union Carbide. Triols with hydroxyl (OH) numbers in the range 23 to 56 and copolymer polyols with a hydroxyl content of 0.90 percent are commercially available. Polymeric MDI with an isocyanate content of about 31 percent and the more volatile tolylyl diisocyanate (TDI) $[(OCN)_2C_6H_3(CH_3)]$ with an isocyanate content of about 87 percent are available.

13.4 Properties of PURs

The properties of some typical PURs are shown in Table 13.1.

Dow supplies DTUL values, at 264 lb/in^2 of 57, 70, 129, and 199°C (135, 158, 265 and 390°F) for its Spectrim structural PUR (SP) and mat-molded (MM) polymers SP 400, SP 500, MM 300, and MM 353, respectively. The DTUL values can be increased by 25°C (45°F) when 50 percent quartz or 20 percent fiberglass is present in PURs.

The abrasion resistance of PUR is related to the ratio of urethane groups to other groups in the polymer molecules. PURs produced from polyethers with terminal hydroxyl groups have superior abrasion re-

TABLE 13.1 Properties of Typical PURs

Property, units*	Elastomeric alloy	Polyester	Polyether
Melting point T_m, °F	329	—	—
Glass transition temp. T_g, °F	—	285	285
Processing temp., °F	400	140	140
Molding pressure, 10^3 lb/in^2	8	1	1
Mold shrinkage, 10^{-3} in/in	2.0	0.01	0.01
Heat deflection temp. under flexural load of 264 lb/in^2, °F	160	165	165
Maximum resistance to continuous heat, °F	150	150	150
Coefficient of linear expansion, 10^{-6} in/(in · °F)	150	150	150
Compressive strength, 10^3 lb/in^2	10	20	20
Izod impact strength, ft · lb/in of notch	No break	No break	No break
Tensile strength, 10^3 lb/in^2	12	5	5
Flexural strength, 10^3 lb/in^2	3	5	5
Percent elongation	350	250	250
Tensile modulus, 10^3 lb/in^2	3	—	—
Flexural modulus, 10^3 lb/in^2	10	—	—
Shore A hardness	70	75	80
Specific gravity	1	1.1	1.1
Percent water absorption	—	—	—
Dielectric constant	—	—	—
Dielectric strength, V/mil	400	470	470
Resistance to chemicals at 75°F:†			
Nonoxidizing acids (20% H_2SO_4)	Q	Q	Q
Oxidizing acids (10% HNO_3)	U	U	U
Aqueous salt solutions (NaCl)	S	S	S
Polar solvents (C_2H_5OH)	U	U	U
Nonpolar solvents (C_6H_6)	Q	Q	Q
Water	S	S	S
Aqueous alkaline solutions (NaOH)	Q	Q	Q

*lb/(in^2 · 0.145) = KPa (kilopascals); ft · lb/(in of notch · 0.0187) = cm · N/cm of notch.
†S = satisfactory; Q = questionable; U = unsatisfactory.

sistance, as measured on a Taber abrader, to those produced from polyester diols. The comparative data are shown in Table 13.2. Electrical properties of PUR are shown in Table 13.3.

13.5 Toxicity of PURs

Most suppliers of diisocyanates and many associations, such as the Manufacturers' Chemists Association, have published recommendations for handling diisocyanates, such as MDI and TDI. MDI has a relatively low vapor pressure at room temperature, and the vapor pressure is much lower for polymeric MCI. All diisocyanates are toxic if inhaled, and the risk is proportional to the vapor pressure, which in-

TABLE 13.2 Abrasion Resistance of PUR and Other Elastomers

PUR	3.0
NBR	42
NR	146
SBR	17
IIR	205
CR	280

TABLE 13.3 Electrical Properties of PUR

Surface resistivity	1.1×10^{17} Ω/square
Volume resistivity	1.0×10^{16} Ω/cm
Arc resistance	67 s
High voltage arc tracking rate	8.6 in/min
High ampere arc ignition	43 arcs
UL electrical properties temperature index	185°F

creases as the temperature is increased. The National Institute of Health and Safety (NIOHS) recommends a time-weighted average (TWA) of the diisocyanate over an 8-h period of 0.02 milligrams per square meter (mg/m^2).

Nevertheless, workers with asthma or bronchitis should not handle MDI or other diisocyanates. These substances may cause coughing, and the skin of workers exposed to diisocyanates may become irritated. Of course, adequate ventilation should always be supplied, and any spilled diisocyanate should be decontaminated by sweeping up with a mixture of fuller's earth, sodium carbonate, and liquid detergent. The decontaminated mixture should be destroyed by incineration.

The polyols are essentially nontoxic. However, care must be taken to protect the worker against the hazards of additives, for example, catalysts, such as tertiary amines (NR_3) and stannous octoate [$SnOOC(CH_2)_6CH_3$], chain extenders, such as 3,3-dichloro-4,4′-diphenylmethane (MOCA), and flame retardants, such as trifluoroethyl phosphate. MOCA is classified as an industrial substance suspected of human carcinogenic potential by the American Conference of Government Hygienists (ACGH) and should be used in a closed system.

In spite of the hazards associated with reactants and additives, PURs are used, without toxicity problems, as spandex swim wear and foundation garments and as artificial organs, including hearts.

13.6 Applications of PURs

In addition to their use for RIM, rigid and flexible foams, spandex snap-back fibers (Lycra, Vyrene), elastomers (Cyanoprene, Estane,

Pellathane, Roylar, Rucothane, and Santoprene), and nonelastic block copolymers (Adiprene, Vulcoprene), PURs are also used as coatings, caulks, and sealants. These compositions may be two-packaged casting systems, blocked PURs, or moisture-cured PURs.

Blocked PURs are PURs produced by the reaction of phenol with isocyanate terminal groups. The phenol or other easily removed hydroxyl compounds are removed when the PUR is heated in the presence of a diol which then reacts with the exposed isocyanate group. PURs with unreacted isocyanate groups react with water to produce carbon dioxide (CO_2) and a diamine [$R(NH_2)_2$], which reacts with isocyanate groups to produce polyurea moieties.

PURs are used as abrasion-resistant coatings, shoe soles, ski boots, catheters, inflatable trousers, and solid and inflatable tires. The production of these long-wearing tires is less labor-intensive than the procedures now used with other polymers, and PURs will replace natural and SBR rubber for this important end use in the near future.

13.7 Isocyanate Polymers and Copolymers

Alkyl isocyanates may be polymerized in the presence of anionic initiators, such as sodium cyanide, to produce polyisocyanates (substituted nylon 1) with relatively high melting points, which decrease as the size of the side chain increases from ethyl [246°C (475°F)] to octadecyl [$C_{18}H_{37}$, 93°C (200°F)]. Copolymers of aryl isocyanates (MDI) may also be produced from ketenes (RC=C=O) and aldehydes (RCHO). PAs (Estamid) are produced by the reaction of diisocyanates and dicarboxylic acids (HOOCRCOOH). PAIs with excellent thermal stability are produced by the reaction of diisocyanates and anhydrides of tricarboxylic acids, such as trimellitic anhydride ($HOOCzC_6H_3C_2O_3$).

PIs (Dow's Polyimide 2080) are produced by the reaction of diisocyanates and the anhydride of a tetracarboxylic acid, such as benzophenone tetracarboxylic acid. Polyheterocyclic film with excellent thermal resistance is produced by the reaction of diisocyanates and hydrogen cyanide (HCN). These polyparabamic acids have been test marketed by Exxon.

13.8 Cyanate Ester Polymers

Cyanate esters, such as BPA diisocyanate, have been produced by the condensation of BPA [$(HOC_6H_4)_2C(CH_3)_2$], cyanogen chloride (ClCN), and triethylamine [$(C_2H_5)_3N$] at low temperatures. Commercial polymers are produced by the cyclotrimerization of these cyanates to (*s*)-

triazine $(ArOCN)_3$. These triazines have been blended with polyethers and polyesters to produce semi-interpenetrating network (IPN) high-modulus polymers.

13.9 Polyureas

Polyureas [-(-RNCONR-)-] are readily formed by the room-temperature reaction of a diisocyanate, such as MDI, and a diamine. Since the presence of moisture causes diisocyanates to decompose to carbon dioxide and a diamine, these polyurea groups are often present in PURs. Alkyl diamines are also added intentionally to the diol mixture in order to produce random or block copolymers with properties that are superior to those of homopolyurethanes. Chain extenders, such as MOCA and other diamines, produce urea groups in the polymer.

13.10 Glossary

ACGH American Conference of Government Hygienists.

Adiprene A PUR elastomer.

alkyl (R) A hydrocarbon group related to natural gas or gasoline.

amine, tertiary R_3N

Ar Aryl.

aryl An aromatic radical related to benzene.

Bayer, O. The inventor of polyurethanes.

bisphenol A $(HOC_6H_4)_2C(CH_3)_2$

blocked polyurethane A phenyl urethane in which the phenol group is readily removed by heating. This should not be confused with a block copolymer which contains PUR and other repeating polymer groups.

carbodiimide -(-RN=C=N-)-

cyanate CNO

cyanogen chloride (ClCN)

Cyanoprene A PUR elastomer.

diamine H_2NRNH_2

dicarboxylic acid HOOCRCOOH

DTUL Heat deflection temperature under load.

Estane A PUR elastomer.

gate An orifice through which molten polymer or reactants (in RIM) enter the mold cavity.

heterocyclic Ring compounds consisting of carbon and other atoms.

IPN Interpenetrating network.

isocyanate NCO

LIM Liquid injection molding.

Lycra A spandex.

MDI Diphenylmethane diisocyanate.

MOCA 3,3-dichloro-4,4′-diphenylmethane.

parabamic acid A prepolymer.

Pellathane A PUR elastomer.

polyol —$R(OH)_n$, where $n = 2$ or more.

polyurea $\text{-(}RNCONR\text{)-}$.

PUR Polyurethane.

R Alkyl.

reaction injection molding (RIM) A process in which the polymer reaction occurs in the mold cavity; also called liquid injection molding (LIM).

RIM Reaction injection molding.

Roylar A PUR elastomer.

Rucothane A PUR elastomer.

Santoprene A PUR elastomer.

Spandex A snap-back filament.

Spectrim A polyurethane.

taber abrader A test instrument in which a coated panel or test specimen revolves in a horizontal position while two weighted abrasive wheels are rotated by friction against the panel on test specimens. The abrasion resistance rating is based on the weight loss after a specificed number of revolutions.

TDI Tolylyl diisocyanate $(OCN)_2C_6H_3(CH_3)$

triazine $(ArOCN)_3$

trichloroethyl phosphate $O{=}P[O(CH_2)_2Cl]_3$

triethylamine $(C_2H_5)_3N$

trimellitic anhydride $HOOCC_6H_3C_2O_3$

triol $R(OH)_3$

TWA Time-weighted average.

urea group —NHCONH—

urethane group —NHCOO—

viscosity Resistance to flow.

Vulcoprene A PUR elastomer.

Vyrene A Spandex.

13.11 References

Becker, W. E.: *Reaction Injection Molding,* Van Nostrand Reinhold, Co., Inc., New York, 1979.

Buist, J. M. (ed.): *Developments in Polyurethanes-1,* Applied Science Publishers, London, 1978.

Donbrow, B. A.: *Polyurethanes,* John Wiley & Sons, Inc. New York, 1968.

Graver, R. B.: Chapter 34 in R. B. Seymour and G. S. Kirshenbaum (eds.): *High Performance Polymers: Their Origin and Development,* Elsevier Science Publishing Co., New York, 1986.

Oertal, G.: *Polyurethane Handbook,* Hanser Publishers, Munich, 1985.

Saunders,J. H., and K. C. Frisch: *Polyurethanes: Their Chemistry and Technology,* Interscience Publishers, John Wiley & Sons, Inc. New York, 1962.

Seymour, R. B.: in *Encyclopedia of Physical Science and Technology,* vol. 2, Academic Press, Inc., Orlando, 1987.

Woods, G.: *The ICI Polyurethane Book*, John Wiley & Sons, Inc. New York, 1987.

Chapter

14

Polyphenylene Oxide (Ether) and Other Polymeric Blends

14.1 Introduction

Blends of gutta percha and hevea rubber were made before Charles Goodyear vulcanized rubber by heating it with sulfur in 1838. Later, reclaimed rubber was blended with virgin rubber in order to reduce costs. Nevertheless, the most widely used commercial blends prior to the 1960s were those of hevea rubber and PS, which were used for the production of HIPS, and those of styrene-AN copolymer and acrylonitrile-butadiene rubber (NBR), which were used for the production of ABS plastics.

While economics was the catalyst for blending reclaimed rubber and virgin natural rubber, there was little economic incentive to blend the inexpensive general-purpose thermoplastics, namely, PE, PVC, and PS, with other polymers.

However, when A. S. Hay produced PPO in 1956, by the copper-catalyzed oxidative coupling of 2,6-dimethylphenol, he was unable to process this engineering resin using standard molding and extrusion techniques. Fortunately, the blend of PS and PPO, which was obtained when the PPO was purged from the processing machinery, proved to be readily moldable.

As might be anticipated, the PS reduced DTUL in accordance with the percentage present in the blend, but the blend with an optimum ratio of PS and PPO became the commercial product sold by General Electric under the trade name Noryl. The successful use of these blends revived a latent interest in mixtures of polymers and blends or alloys, and these have become important commercial plastics.

14.2 Blends

Many natural products, such as wood, are blends of more than one polymer. However, until recently, commercial blends were restricted to those based on economics, such as blends of virgin NR and reclaimed rubber, those based on material shortages, such as blends of NR and SR, and those in which improvement in physical properties was readily demonstrated, such as MacIntosh rubber-cloth sandwich, and fiber blends, such as cotton and polyester.

However, the success of PPO-PS blends (Noryl) and the ease of production of blends catalyzed a renewed interest in polymer blends or alloys. Since Noryl was produced by General Electric, it is not surprising to note that General Electric also produces Xenoy, which is a blend of PC and PET and Noryl GTX, which is a blend of PPO and nylon. Upjohn also produces a blend of PI and polyphenylene sulfide (PPS; Upjohn 20).

The term "alloy" is used to describe miscible or immiscible mixtures of polymers, which are usually blended as melts. Miscible blends are characterized by properties, such as T_g values, which are between those of the components. The miscibility may be the result of similar solubility parameters, strong intermolecular attractions, such as hydrogen bonding, or the presence of a compatibilizing agent. Immiscible blends are characterized by two or more phases, i.e., continuous and discontinuous phases, and more than one T_g value. The ultimate in blends are block copolymers, such as styrene-isoprene-styrene block copolymers, which are sold by Shell under the trade name Kraton.

However, the most widely used blend is the blend of PPO and PS (usually HIPS), which is one of the "big five" engineering thermoplastics. The T_g value of this miscible modified PPO is between the T_g of PPO [208°C (406°F)] and the T_g of PS [100°C (212°F)]. In 1988, 82,000 t of modified PPO (also called polyphenylene ether) was produced in the United States. As evident from the index, the properties of many other blends are discussed throughout this book.

14.3 Blending Concepts

R. Mendelsohn has classified blends and alloys as miscible one-phase systems and partially miscible or immiscible two-phase systems. The Versailles Project on Advanced Materials (VANAS) has been unsuccessful in its attempts to establish standards for these important modern materials.

A negative change in free energy ($-\Delta G$) is required for miscibility. The change in entropy (ΔS) is essentially negligible, and hence the change in enthalpy (ΔH) must be negative or zero for the formation of miscible blends as shown by the Gibbs free energy equation:

$$\Delta G = \Delta H - T\,\Delta S$$

where T is the absolute temperature.

This requirement for a negative ΔH value can be met when there is a physical attraction, such as hydrogen bonding, between the component polymers. The T_g and many properties of miscible blends can be predicted from the law of additives. Thus, the heat deflection point of the compatible blends of PPO and PS (Noryl) can be predicted from information on their composition. The DTUL values of Noryl vary from 91 to 149°C (195 to 300°F) and are inversely related to the PS content. This statement also applies to Prevex, which is a blend of PS and polyphenylene ether (PPE), and which, like Noryl, is available from General Electric.

Blends of PC and PBT (Xenoy) are partially miscible, but blends of PC and a copolyester of hexane dimethanol and isophthalic and terephthalic acids (Kodar A) are miscible and have only one T_g value. Blends of high-molecular-weight polymers usually exhibit a low critical solution temperature, while blends of low-molecular-weight polymers usually exhibit an upper critical solution temperature. The properties and blends of PVC and aliphatic polyesters with and without pendant groups are similar. However, polyesters with pendant groups are immiscible with PC.

While Noryl and Xenoy are miscible blends, most polymer blends are immiscible. The dispersed phase in these blends may be present as spheres, fibrils, or platelets. Block copolymers or chlorinated PE may be added as compatibilizing agents.

A sensitive component may be protected from degradation in a blend. For example, while PMMA is degraded by gamma radiation, the rate of the degradation is reduced in a PMMA-SAN blend. The blending of an amorphous polymer, such as PMMA, with a crystalline polymer, such as PVDF, suppresses crystallinity, and the blend is amorphous when the PVDF content is less than 40 percent.

The modulus, tensile strength, and heat deflection temperature of an immiscible blend are related to the properties of the principal component of the blend. Stress transfer from spheres in an immiscible blend may be facilitated by changing the morphology to lamellar, or, as in the case with fiber-reinforced composites, rubbery properties of a blend may be enhanced by good interfacial adhesion between the components.

Injection-moldable parts may have a weld line which reduces ductility of the molded part. This deficiency may be overcome by adding a small amount of compatibilizer. Blends with lamellar morphology, such as nylon-HDPE or PET–ethylene-co-vinyl alcohol (EVAL), exhibit resistance to moisture permeation. The impact resistance of rubber-modified (EPDM) nylon like HIPS is superior to that of the unblended nylon. This improvement in toughness may depend on energy dissipation resulting from debonding at the interface during fracture.

14.4 Properties of Commercial Polymer Blends

The properties of PPO-PS blends are superior to those of the components present in the blend. The energy of compounding PPO blends has been reduced by blending with the zinc salt of sulfonated EPDM, the zinc salt of sulfonated PS, zinc stearate, and tricresyl phosphate. Noryl, Prevex, and other blends are being blow-molded to produce automobile bumpers and furniture.

The production of nylon has been improved by blending with PPO (Vydyne, Noryl GTX). Nylon also has been blended with PC; HDPE (Selar); PP, SAN, and ABS (Triax, Elemid); PBT (Bexloy); and polyarylates (Bexlar). These blends have lower water absorption than nylon 6,6. Nylon-SAN, which has a high impact resistance of 16 ft · lb/in of notch, has received a UL rating of 104°C (220°F). The nylon-arylate blend is transparent and has a DTUL value of 154°C (310°F).

The addition of SMA-AN terpolymer (Cadon) to PVC increases the DTUL value and the resistance to impact to 100°C (212°F) and 5.6 ft · lb/in of notch, respectively. The addition of SMA also increases the temperature resistance of PVC.

The addition of ABS to PC (Proloy, Bayblend) lowers the DTUL and resistance to impact to 116°C (240°F) and 10.5 ft · lb/in of notch, respectively. Likewise, the addition of PBT to PC (Makroblend, Xenoy) reduces the DTUL value and the resistance to impact to 79°C (175°F) and 14 ft · lb/in of notch, respectively.

PC is also being blended with SMA copolymer (Arloy), thermoplastic urethane (Texin), and PE (Lexan).

The properties of several commercial polymer blends are shown in Table 14.1.

14.5 Modified PPO

PPO, which has the lowest water absorption of any engineering plastic, is an amorphous resin which is used primarily as an injection-molded plastic. It is also used as plastic sheet for thermoforming. PPO has inherent stiffness based on the phenylene group, and a small degree of flexibility based on the presence of the ether group in the polymer chain (see the structural formula for its repeating unit in Fig. 14.1).

TABLE 14.1 Properties of Typical Polymer Blends

Property, units*	Impact-modified POM	PVC-acrylic alloy	Nylon 6,6-EPDM	PC-PBT	SAN-EPDM	SMA-PC
Melting point T_m, °F	330	—	480	—	—	—
Glass transition temp. T_g, °F	—	220	—	—	120	—
Processing temp., °F	400	—	550	500	450	550
Molding pressure, 10^3 lb/in^2	15	--	15	15	1	--
Mold shrinkage, 10^{-3} in/in	15	—	15	7	6	6
Heat deflection temp. under flexural load of 264 lb/in^2, °F	200	160	160	220	200	230
Maximum resistance to continuous heat, °F	175	135	135	190	175	200
Coefficient of linear expansion, 10^{-6} in/(in · °F)	70	—	75	50	50	—
Compressive strength, 10^3 lb/in^2	—	3	4	7	—	—
Izod impact strength, ft · lb/in of notch	5	6.5	No break	10	15	10
Tensile strength, 10^3 lb/in^2	7.5	10	8	8	5	14
Flexural strength, 10^3 lb/in^2	13	10	1	11	8	12
Percent elongation	125	100	125	140	20	50
Tensile modulus, 10^3 lb/in^2	400	330	—	—	—	—
Flexural modulus, 10^3 lb/in^2	—	330	100	300	280	400
Rockwell hardness	M65	M100	R115	R116	R100	—
Specific gravity	1.35	—	1.15	1.02	1.02	1.15
Percent water absorption	0.2	0.05	1.5	0.15	0.1	0.2
Dielectric constant	—	—	—	—	—	—
Dielectric strength, V/mil	450	400	450	475	420	—
Resistance to chemicals at 75°F:†						
Nonoxidizing acids (20% H_2SO_4)	U	S	U	Q	S	S
Oxidizing acids (10% HNO_3)	U	Q	U	U	U	U
Aqueous salt solutions (NaCl)	S	S	S	S	S	S
Aqueous alkaline solutions (NaOH)	S	Q	Q	U	S	Q
Polar solvents (C_2H_5OH)	S	S	S	S	S	S
Nonpolar solvents (C_6H_6)	Q	U	Q	U	U	U
Water	S	S	S	S	S	S

*lb/(in^2 · 0.145) = KPa (kilopascals); ft · lb/(in of notch · 0.0187) = cm · N/cm of notch.
†S = satisfactory; Q = questionable; U = unsatisfactory.

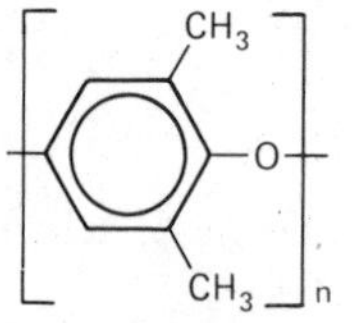

Figure 14.1 PPO.

The flame resistance of PPO depends on the thickness of the sample. UL 94 ratings of HB, V-0, V-1, and V-2 have been assigned for thicknesses as low as 0.06 in. Unpigmented PPO is approved by the USDA for continuous food contact.

Modified PPO is soluble or swollen in solvents such as ammonia, chlorinated solvents, esters, gasoline, and ketones. Modified PPO has been assigned continuous-use temperatures of 49 to 110°C (120 to 230°F) by UL depending on the ratio of PS and PPO in the blend. PPO has moderate resistance to weathering and high-energy radiation.

14.6 Design of PPO Molded Parts

It is advantageous to use PPO and other engineering polymers in the region in which the response of the material is elastic. The designer should note the variations present in different grades of the same engineering resin and make use of the supplier's technical service personnel. It is important to note that the properties reported in published tables have usually been obtained on compression-molded samples, which tend to be more isotropic than samples obtained by extrusion or injection molding.

As is the case with most engineering plastics, one should use gradual rather than sharp transitions in the thickness of molded parts. The land length of gates should be short to ensure uniform flow and ready degating. Vent slots of at least 0.01 and 0.0005 in should be present for reinforced and nonreinforced PPO, respectively. The inclusion of ribs will improve the rigidity of the part, and undercuts should be avoided whenever possible to ensure retaining the integrity of the part during ejection.

Since the high shear modulus of PPO is maintained, even at a relative humidity of 50 percent, up to 160°C (320°F), automotive parts may be painted at elevated temperatures. Tough blends of PPO and nylon (Noryl GTX) are preferred for painted parts. The water absorption of Noryl GTX is a function of the nylon content.

14.7 Properties of PPO-PS Blends

Since General Electric offers 20 grades of Noryl, it is difficult to test the properties of all these grades. Hence, the properties of a few selected grades are shown in Table 14.2.

TABLE 14.2 Properties of Typical Modified PPO

Property, units*	High-PS blend	Low-PS blend	PA blend	PPO-PS + 30% fiberglass
Melting point T_m, °F	—	—	—	—
Glass transition temp. T_g, °F	105	130	135	120
Processing temp., °F	500	500	500	500
Molding pressure, 10^3 lb/in^2	15	15	12	25
Mold shrinkage, 10^{-3} in/in	6	7	6	3
Heat deflection temp. under flexural load of 264 lb/in^2, °F	200	245	240	290
Maximum resistance to continuous heat, °F	180	230	220	265
Coefficient of linear expansion, 10^{-6} in/(in · °F)	25	25	—	12
Compressive strength, 10^3 lb/in^2	14	16	—	18
Izod impact strength, ft · lb/in of notch	5	5	7	20
Tensile strength, 10^3 lb/in^2	8	10	7	17
Flexural strength, 10^3 lb/in^2	10	11	10	20
Percent elongation	50	60	—	3
Tensile modulus, 10^3 lb/in^2	350	365	350	1200
Flexural modulus, 10^3 lb/in^2	350	350	325	1100
Rockwell hardness	R115	R120	M93	R115
Specific gravity	1.08	1.07	1.3	1.3
Percent water absorption	0.08	0.1	0.1	0.08
Dielectric constant	2	2	2	2
Dielectric strength, V/mil	200	200	200	200
Resistance to chemicals at 75°F:†				
Nonoxidizing acids (20% H_2SO_4)	S	S	Q	S
Oxidizing acids (10% HNO_3)	Q	Q	Q	Q
Aqueous salt solutions (NaCl)	S	S	S	S
Polar solvents (C_2H_5OH)	S	S	S	S
Nonpolar solvents (C_6H_6)	U	U	Q	U
Water	S	S	S	S
Aqueous alkaline solutions (NaOH)	S	S	S	S

*lb/(in^2 · 0.145) = KPa (kilopascals); ft · lb/(in of notch · 0.0187) = cm · N/cm of notch.
†S = satisfactory; Q = questionable; U = unsatisfactory.

14.8 PPE

Another PPO, called PPE (Prevex), is produced by the oxidative coupling of a mixture of 2,6-dimethylphenol and 2,3,6-trimethylphenol and was formerly sold by Borg-Warner. This stiff polymer, like Noryl, is available from General Electric. It is usually modified by blending with PS. Nylon-modified blends, like those of Noryl, have good impact resistance, and this property is retained at low temperatures [−40°C (−40°F)]. PPE, like Noryl, is readily injection-molded and injection-extruded.

14.9 Applications of PPO Blends

Several of the commercial unfilled, filled, and flame-retarded grades of Noryl are used for automotive applications, business and computer equipment, electronic applications, household appliances, hair drier housings, and pump housings and impellers.

14.10 Electrical Properties

Noryl and Prevex are characterized by high dielectric strength and low dissipation factors. The electrical properties of a typical PPO blend are shown in Table 14.3.

TABLE 14.3 Electrical Properties of a Typical PPO Blend

Dielectric strength	192 V/mil
Dielectric constant	
At 100 Hz	2.27
At 1,000,000 Hz	2.18
Dissipation factor	
At 100 Hz	4.7×10^3
At 1,000,000 Hz	3.9×10^3

14.11 PC-PBT (Xenoy)

Blends of PC and PBT are available as both reinforced and unreinforced plastics and are sold by General Electric under the trade name Xenoy. This nonmiscible blend is more resistant to gasoline spillage than PC. Blends of PC and the copolymer of hexane dimethanol and terephthalic and isophthalic acids (Kodar A) are compatible, and the moldings are transparent. The properties of impact-modified bumper-grade Xenoy are shown in Table 14.4.

TABLE 14.4 Properties of Typical Impact-Modified PC-PBT Blends

Property, units*	PC-PBT blend	PC-PBT blend + 30% fiberglass
Melting point T_m, °F	—	—
Glass transition temp. T_g, °F	—	—
Processing temp., °F	525	550
Molding pressure, 10^3 lb/in^2	15	20
Mold shrinkage, 10^{-3} in/in	7	6
Heat deflection temp. under flexural load of 264 lb/in^2, °F	240	400
Maximum resistance to continuous heat, °F	220	360
Coefficient of linear expansion, 10^{-6} in/(in · °F)	50	15
Compressive strength, 10^3 lb/in^2	12	11
Izod impact strength, ft · lb/in of notch	10	3
Tensile strength, 10^3 lb/in^2	8	13
Flexural strength, 10^3 lb/in^2	7	20
Percent elongation	150	4
Tensile modulus, 10^3 lb/in^2	300	—
Flexural modulus, 10^3 lb/in^2	275	850
Rockwell hardness	M85	R110
Specific gravity	1.2	1.5
Percent water absorption	0.2	0.1
Dielectric constant	4	4
Dielectric strength, V/mil	500	500
Resistance to chemicals at 75°F:†		
Nonoxidizing acids (20% H_2SO_4)	Q	Q
Oxidizing acids (10% HNO_3)	U	U
Aqueous salt solutions (NaCl)	S	S
Polar solvents (C_2H_5OH)	S	S
Nonpolar solvents (C_6H_6)	Q	Q
Water	S	S
Aqueous alkaline solutions (NaOH)	Q	Q

*lb/(in^2 · 0.145) = KPa (kilopascals); ft · lb/(in of notch · 0.0187) = cm · N/cm of notch.
†S = satisfactory; Q = questionable; U = unsatisfactory.

14.12 PPO-Nylon Blends

Blends of PPO and nylon (PA) are incompatible, but good properties are obtainable through the use of compatibilizing agents. The PPO is dispersed in a continuous nylon matrix in these blends. Because of the incompatibility of the two phases, the modulus decreases very little at the T_g of the PA [71°C (160°F)] and is maintained up to the plasticization of the PPO phase [199°C (390°F)].

This blend, which is sold by General Electric under the trade name Noryl GTX, can be painted and baked at 191°C (375°F) without noticeable warpage or distortion. The properties of these blends are shown in Table 14.5.

TABLE 14.5 Properties of Typical PPO-PA Blends (Noryl GTX 810)

Property, units*	
Melting point T_m, °F	—
Glass transition temp. T_g, °F	—
Processing temp., °F	500
Molding pressure, 10^3 lb/in^2	15
Mold shrinkage, 10^{-3} in/in	10
Heat deflection temp. under flexural load of 264 lb/in^2, °F	370
Maximum resistance to continuous heat, °F	345
Coefficient of linear expansion, 10^{-6} in/(in · °F)	25
Compressive strength, 10^3 lb/in^2	17
Izod impact strength, ft · lb/in of notch	1.5
Tensile strength, 10^3 lb/in^2	13
Flexural strength, 10^3 lb/in^2	19
Percent elongation	10
Tensile modulus, 10^3 lb/in^2	200
Flexural modulus, 10^3 lb/in^2	225
Rockwell hardness	R119
Specific gravity	1.16
Percent water absorption	0.5
Dielectric constant	4
Dielectric strength, V/mil	500
Resistance to chemicals at 75°F:†	
Nonoxidizing acids (20% H_2SO_4)	Q
Oxidizing acids (10% HNO_3)	U
Aqueous salt solutions (NaCl)	S
Polar solvents (C_2H_5OH)	S
Nonpolar solvents (C_6H_6)	Q
Water	S
Aqueous alkaline solutions (NaOH)	S

*lb/(in^2 · 0.145) = KPa (kilopascals); ft · lb/(in of notch · 0.0187) = cm · N/cm of notch.
†S = satisfactory; Q = questionable; U = unsatisfactory.

14.13 Commercial Blends

According to C. H. Kline, the U.S. specialty-plastics blend and alloy market will approach $1 billion. Kline states that PPO-PS blends (Noryl), impact-modified nylons, and ABS-PVC blends will continue to dominate plastic-blend market sales. The trade names, compositions, and producing companies of the leading polymeric blends are shown below:

Trade name	Composition	Producing company
Elemid	ABS-nylon	Borg-Warner (GE)
Triax	ABS-nylon	Monsanto
Proloy	PC-ABS	Borg-Warner (GE)
Pulse	PC-ABS	Dow
Texin	PC-TPE	Mobay

Trade name	Composition	Producing company
Tarblend	PC-ASA	BASF
Duraloy	POM-HDPE	Hoechst Celanese
Noryl GTX	Nylon-PPO	General Electric
Bexloy	Nylon-PBT	Mobay

Mobay maintains that its PC-PBT blend is a serious contender for many end-use applications and predicts that the U.S. sales of this blend will be 7000 to 9000 t in the early 1990s. Nylon-PC blend (Dexcarb) and Nylon-PP blend (Dexlon) are heat-resistant, injection-moldable, impact-resistant polymers with many potential applications.

14.14 Outlook for Blends

Because of the ease of production of blends and alloys on existing equipment, the growth of these high-performance polymers will be in the order of 40 percent in the next decade (1990 to 2000). Blends account for almost 20 percent of the high-performance thermoplastics marketed at present, and they will retain this percentage.

The worldwide production of high-performance polymers will exceed 2,000,000 t in 1991. The United States, with an annual production of 810,000 t, will continue to lead, but it will be followed closely by western Europe and Japan with annual volumes of 720,000 and 460,000 t, respectively.

GM is producing automobile fenders from PPO-nylon alloys, and the shift from metal to all-plastic automobile bodies will be a reality in the year 2000. Blends of nylon and PBT (Bexloy) are also being used in many automotive applications. High-speed stirring techniques have been used to compatibilize immiscible blends.

Computer-aided manufacturing (CAM) will catalyze the increased use of polymer blends. Savings of $30,000 for each pound reduction in weight of aerospace vehicles favors the use of high-performance polymers. The development of new high-performance polymers in Japan, West Germany, and the United States will catalyze the use of all types of polymers in a wide variety of applications and open up new opportunities for uses in which excellent performance and freedom of design are essential.

14.15 Glossary

ABS A terpolymer of acrylonitrile, butadiene, and styrene.

agent, compatibilizing An additive such as chlorosulfonated polyethylene which makes incompatible polymers more compatible.

aliphatic A straight-chained molecule, like gasoline.

alloy A blend of polymers.

ammonia NH_3

amorphous Noncrystalline.

arc resistance The ability of a plastic to withstand exposure to voltage.

Arloy A PC-SMA blend.

ASA Acrylonitrile-styrene-ethyl acrylate terpolymer.

Bayblend An ABS-PC blend.

Bexlar A nylon-polyarylate blend.

Bexloy A nylon-PBT blend.

Cadon A terpolymer of styrene–maleic anhydride–acrylonitrile.

CAM Computer-aided manufacturing.

chlorinated solvents RCl

copolymer, block A copolymer that consists of long sequences of polymers of different monomers in the principal chain.

ΔG The change in Gibbs free energy.

ΔH The change in enthalpy (heat content).

ΔS The change in entropy (disorder).

Dexcarb A nylon-PC blend.

Dextron A nylon-PP blend.

dielectric constant The ratio of the capacitance of a capacitor containing the test material and that of a capacitor containing a vacuum or air.

dielectric strength The maximum electric potential gradient that the polymer can withstand without rupture.

dissipation A loss of energy.

dissipation factor The ratio of the power loss to the total power transmitted through a plastic.

DTUL Deflection temperature under load, usually 264 lb/in^2.

Duraloy A POM-HDPE blend.

Elemid A nylon-ABS blend.

EPDM Ethylene-propylene copolymer.

ester RCOOR

EVAL Polyethylene-co-vinyl alcohol: $-(CH_2CH_2CH_2CHOH)-$

flame resistance These data are restricted to a specific test and should be used as guidelines only.

Gibbs free energy equation $\Delta G = \Delta H - T\,\Delta S$

glass transition temperature The temperature at which a hard brittle polymer becomes flexible as the temperature is increased.

gutta percha A hard naturally occurring trans polyisoprene.

hevea rubber Natural rubber.

high-impact polystyrene Rubber-toughened polystyrene.

HIPS High-impact polystyrene.

hydrogen bond The attraction between a hydrogen atom and an oxygen or nitrogen atom in an organic compound.

ketones $RC(=O)R$

Kodar A A copolymer of hexane dimethanol and isophthalic and terephthalic acids.

Kraton A block copolymer of styrene-isoprene-styrene.

Lexan A PC-HDPE blend.

Makroblend A PC-PBT blend.

matrix The continuous phase; resin.

morphology Shape.

NBR A copolymer of acrylonitrile and butadiene.

Noryl A modified PPO.

Noryl GTX A blend of PPO and nylon.

NR Natural rubber.

Nylon 6,6 $\text{-[}NH(CH_2)_6NHCO(CH_2)_4CO\text{]-}$

PA Polyamide; nylon.

PBT Polybutylene terephthalate.

PC Polycarbonate.

pendant group A functional group on the principal polymer chain.

PET Polyethylene terephthalate.

PI Polyimide.

PMMA Polymethyl methacrylate: $[CH_2CH(CH_3)COOCH_3—]$

polyphenylene oxide PPO, PPE: $\left[-C_6H_2(CH_3)_2-O- \right]$ (2,6-dimethyl-1,4-phenylene oxide repeat unit, with CH_3 groups on the ring)

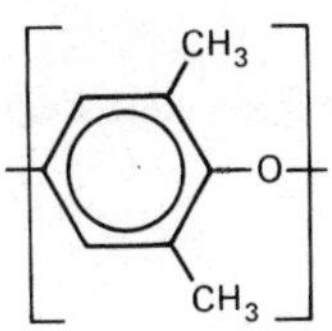

POM Acetal polymer or copolymer.

PPE Polyphenylene ether, a term used to describe the oxidative-coupled copolymer of dimethyl- and trimethylphenol.

PPO Polyphenylene oxide.

PPS Polyphenylene sulfide.

Prevex Polyphenylene ether.

Proloy An ABS-PC blend.

PS Polystyrene.

Pulse A PC-ABS blend.

PUR Polyurethane.

PVC Polyvinyl chloride.

PVDF Polyvinylidene fluoride; Kynar; $-(CH_2CF_2)-$.

rubber, reclaimed Recycled rubber.

SAN Styrene-acrylonitrile copolymer.

Selar A nylon-HDPE blend.

shear stress The stress between sliding surfaces of two planes.

SMA Styrene–maleic anhydride copolymer.

SR Synthetic rubber.

surface resistivity The electric resistance measured on opposite sides of a square.

T Degrees K (kelvin).

Tarblend A PC-ASA blend.

Texan A PC-PUR blend.

T_g Glass transition temperature.

TPE Thermoplastic elastomer.

Triax A nylon-ABS blend.

UL Underwriters' Laboratory.

USDA U.S. Department of Agriculture.

volume resistivity The ratio of the potential gradient parallel to the current in the plastic to the current density.

Xenoy A blend of PC and PBT.

14.16 References

Feth, G. E.: Chapter 5 in J. Margolis (ed.): *Engineering Thermoplastics,* Marcel Dekker, Inc., New York, 1985.

Hay, A. S., and J. M. Heuschen: in "Noryl Resins," R. B. Seymour and G. S. Kirshenbaum (eds.): *High Performance Polymers: Their Origin and Development,* Elsevier Science Publishing Co., Inc., New York, 1986.

Heuschen, J. M., A. S. Hoy, and L. H. Sperling: in "Thermoplastic Blends," R. B. Seymour and G. S. Kirshenbaum (eds.): *High Performance Polymers: Their Origin and Development,* Elsevier Science Publishing Co., Inc., New York, 1986.

Manson, J. A., and L. H. Sperling: *Polymer Blends and Composites,* Plenum Press, Plenum Publishing Corporation, New York, 1976.

Olabisi, O., L. M. Robeson, and M. T. Snow: *Polymer-Polymer Miscibility,* Academic Press, Inc., New York, 1979.

Paul, D. R., and S. Neuman: *Polymer Blends,* vol. 2, Academic Press, Inc., New York, 1978.

Chapter 15

Fluorocarbon Polymers

15.1 Introduction

Fluorine (F) and inorganic fluorides have been known for many years, but organic fluorides are relatively new chemicals. Swarts, who was the father-in-law of Leo Baekeland, produced organic fluorine compounds by heating organic chlorine compounds with hydrogen fluoride (HF) in the presence of antimony fluoride (SbF_3)in the late 1800s.

The commercialization of volatile organic fluoride compounds was undertaken by Kinetic Chemical (a joint venture of GM and Du Pont) in order to provide a nontoxic, nonflammable refrigerant which could be used in place of ammonia (NH_3), sulfur dioxide (SO_2), and volatile aliphatic hydrocarbons, such as ethylene ($H_2C{=}CH_2$). Dichlorodifluoromethane (CCl_2F_2), which was the first fluorocarbon refrigerant, was introduced under the trade name Freon in 1931. Subsequently, these colorless, nonflammable, nontoxic, odorless compounds were widely used as aerosol propellants, but their use is being reduced because of their alleged effect on the ozone layer in the outer atmosphere.

In 1934, Henne of Ohio State University produced TFE ($F_2C{=}CF_2$), and in 1936, Plunkett of Du Pont chlorinated this TFE in an attempt to obtain a superior Freon-type refrigerant. The discovery of a solid polymer (Teflon: PTFE) in one of the cylinders by Plunkett and his assistant, Jack Rebok, has been reported many times and was highly publicized by Du Pont in 1988 on the occasion of the fiftieth birthday of Teflon.

The discovery of PTFE has been called serendipity by some, since the goal of Plunkett, like that of the three princes of Serendip (Sri Lanka), was different than what he discovered. However, it should be noted that most of Plunkett's contemporaries believed that it is not possible to polymerize a fluorinated vinyl monomer, and many outstanding organic chemists have discarded polymers which were produced accidentally. In any case, the PTFE patent was granted to Plunkett, and this discovery showed that Louis Pasteur was correct when he said, "Chance favors the prepared mind."

The PTFE from Du Pont's pilot plant was utilized immediately for gaskets and packaging in the separation of the isotopes of uranium hexafluoride at Oak Ridge, Tennessee. In addition to being essential for the development of the first atomic bomb, PTFE was used as a nose cone cover and for proximity fuses on artillery shells during World War II. PTFE not only continues to be the major polyfluorocarbon, but its discovery set the stage for the synthesis of many other polyfluorocarbons.

15.2 Structure of PTFE

Since TFE is readily polymerized by peroxy free radicals in aqueous media, it is surprising that PTFE was not discovered earlier by those who were convinced that TFE is not polymerizable. PTFE [$\leftarrow CF_2CF_2 \rightarrow$] is a highly crystalline polymer in which the fluorine pendant groups are closely packed on the polymer backbone. The remarkable properties of PTFE are related to the close packing and strong C—F and C—C covalent bonds present in PTFE.

PTFE is an extremely high molecular weight thermoplastic, but it is not processible by conventional injection- or compression-molding or extrusion techniques. Hence, it is fabricated by sintering, like metal powders. Thus, a billet is produced by cold pressing finely divided PTFE particles, and this billet or preform is coalesced by heating at 327°C (620°F). The white-bluish-gray, translucent waxy billet is then fabricated by machining.

Coagulated PTFE may be blended with 40 volume percent kerosene, compressed into a billet, and extruded through an orifice as sheet, filament, or wire coating. The cross section of the extrudate is controlled by the dimensions of the die, but the length of the profile is increased severalfold over the normal length. As a result, the extrudate is porous and has a melting point that is 78°C (140°F) higher than that of the unextruded PTFE.

Because of the difficulty of processing the highly crystalline homopolymer, copolymers of TFE and a small amount of hexafluoropropylene (HFP) are often used. In addition to the trade name Teflon used by Du Pont, Allied-Signal and ICI use the trade names Halon and Fluon respectively.

15.3 Properties of PTFE

The properties of PTFE and fiberglass-filled PTFE are shown in Table 15.1.

TABLE 15.1 Properties of Typical Polyfluorocarbons

Property, units*	PTFE	PTFE + 25% fiber-glass	PCTFE	PVDF	PVF
Melting point T_m, °F	620	620	428	347	—
Glass transition temp. T_g, °F	—	—	—	—	22
Processing temp., °F	600	600	500	400	400
Molding pressure, 10^3 lb/in^2	4	5	4	3	3
Mold shrinkage, 10^{-3} in/in	30	20	12	20	20
Heat deflection temp. under flexural load of 264 lb/in^2, °F	210	220	—	210	210
Maximum resistance to continuous heat, °F	500	500	--	200	275
Coefficient of linear expansion, 10^{-6} in/(in · °F)	40	40	35	40	50
Compressive strength, 10^3 lb/in^2	17	12	60	12	—
Izod impact strength, ft · lb/in of notch	3	2.7	4	3	2
Tensile strength, 10^3 lb/in^2	3	2	5	5	30
Flexural strength, 10^3 lb/in^2	—	2	9	50	150
Percent elongation	300	250	175	50	150
Tensile modulus, 10^3 lb/in^2	70	220	225	300	200
Flexural modulus, 10^3 lb/in^2	80	200	225	350	200
Shore D hardness	60	65	80	80	64
Specific gravity	2.2	2.2	2.1	1.8	1.4
Percent water absorption	0.01	0.01	0	0.04	0
Dielectric constant	2.5	2.5	2.5	8	8
Dielectric strength, V/mil	480	320	550	270	200
Resistance to chemicals at 75°F:†					
Nonoxidizing acids (20% H_2SO_4)	S	S	S	S	S
Oxidizing acids (10% HNO_3)	S	S	S	S	S
Aqueous salt solutions (NaCl)	S	S	S	S	S
Polar solvents (C_2H_5OH)	S	S	S	Q	S
Nonpolar solvents (C_6H_6)	S	S	S	S	S
Water	S	S	S	S	S
Aqueous alkaline solutions (NaOH)	S	S	S	S	S

*lb/(in^2 · 0.145) = KPa (kilopascals); ft · lb/(in of notch · 0.0187) = cm · N/cm of notch.
†S = satisfactory; Q = questionable; U = unsatisfactory.

15.4 Frictional Behavior of Polyfluorocarbons

A high coefficient of friction, μ, is required for elastomers used in tire tread, but a low value is essential when a polymer, such as PTFE, is used as a bearing. The coefficient of friction between moving surfaces is equal to the ratio of the tangential force F and the load W, i.e., $\mu = F/W$.

The coefficient of friction is independent of the contact area and the sliding velocity but is related to molecular adhesion and hardness of the materials involved. The exceptionally low coefficient of friction of PTFE is a function of its molecular cohesion and low surface energy, which are related to the close packing of the fluorine pendant groups, which provide a smooth molecular profile for the polymer chain. Thus, this coefficient increases as the fluorine atom is replaced by chlorine or hydrogen atoms, i.e., as one goes from PTFE to PCTFE to PVDF and to PVF.

As shown by the data in Table 15.2, the coefficient of friction, measured at low speed, for polymers against metal is 0.04 for PTFE and 2.0 for NR.

TABLE 15.2 Coefficient of Friction of Polymers Against Metal and Against Polymers

Polymer	μ
PTFE	0.04
HDPE	0.23
Nylon 6,6	0.36
PCTFE	0.56
PVC	0.6
NR	2.0

15.5 Abrasion Resistance of Polyfluorocarbons

Abrasion resistance, which is related to friction and hardness, is usually measured by weight loss during the abrasion process. Scratch resistance is related to abrasion, and there is a correlation between the rate of abrasive wear and the reciprocal of the product of the tensile strength and elongation. The abrasion rate measured by a bronze ball sliding over the polymer surface is shown in Table 15.3.

15.6 PCTFE

Low-molecular-weight polymers of chlorotrifluoroethylene (CTFE) were produced by chemists at IG Farbenindustrie in 1957. Subse-

TABLE 15.3 Relative Abrasion Rates of Polymers

	Unfilled	30% fiberglass-filled	PTFE with other fillers (30%)
Nylon 6,6	< 0.01	7.3	
PET	< 0.01	146	
POM	< 0.01	15	
PTFE	< 0.01	62	
Graphite			0.05
Powder			0.08
Asbestos			10
Carbon fiber			8
Mica			3.0

quently, somewhat higher molecular weight PCTFE was produced at Cornell University and commercialized for use as a lubricant by Hooker Electrochemical. Much higher molecular weight PTFE and copolymers of CTFE and vinylidine fluoride (VDF) were produced by M. W. Kellogg, under the trade name Kel-F.

Since the presence of a pendant chlorine group in the monomer ($F_2C{=}CFCl$) reduces crystallinity and rigidity, PCTFE is a syndiotactic polymer which can be injection-molded and extruded by conventional techniques. This improvement in processibility is paralleled by a moderate decrease in resistance to corrosives and solvents. The copolymer with 25 percent VDF is a soft resin, and the copolymer with 50 percent VDF is an elastomer. The permeability of this relatively clear amorphous PCTFE to water vapor is much lower than that of Saran, nylon, or PET.

15.7 PVDF

PVDF was patented by T. Ford and W. Hanford in 1948 and was introduced commercially by Pennwalt under the trade name Kynar, in 1961. It is also available under the trade name Foroflon from Atochem. This crystalline polymer is produced by the free-radical emulsion polymerization of VDF ($H_2C{=}CF_2$).

PVDF may be injection-molded and extruded by conventional techniques. In addition to its excellent electrical and corrosion-resistance properties, PVDF can be used as a heat-shrinkable tubing, conduit-free plenum cable, microporous membrane, and metal coating. PVDF is insoluble in nonpolar solvents but is soluble in strong polar solvents and alkyl amides.

PVDF is classified by UL 910 as an insulator. PVDF film has piezoelectrical properties, i.e., it changes its polarization in response to mechanical stress. The properties of PVDF are shown in Table 15.1.

PVDF is much more expensive than general-purpose polymers, such as HDPE, but it is the least expensive of the polyfluorocarbons. Costs are also reduced by bonding PVDF film to steel or plastic laminates.

15.8 PVF

Vinyl fluoride ($H_2C{=}CHF$) was synthesized by Swarts in 1901, but PVF, a high-melting-point, transparent chemical-resistant polymer [$\text{-(}CH_2CHF\text{)-}$], was not produced commercially until the early 1940s. PVF is a crystalline polymer, which is used primarily as a film (Tedlar), which is usually laminated to metal or other substrates to provide a weather-resistant laminate. The properties of PVF are shown in Table 5.1.

15.9 Fluorocarbon Copolymers

As discussed in Sec. 6.9, copolymers of VDF and CTFE (Kel F), and of VDF and hexafluoropropylene [$F_2C{=}CF(CF_3)$] and other fluoro monomers have been used as engineering elastomers. As shown by the data in Table 15.4, the T_g values of the CTFE-VDF copolymers decrease as the percentage of VDF increases.

In 1962, S. Dixon, D. R. Rexford, and J. S. Rugg patented the copolymer of VDF and HFP (Viton A, Fluorel), and in 1961, J. E. Pailthorp and H. E. Schroeder patented the terpolymer of VDF, HFP, and TFE (Viton B). These copolymers, like Kel-F elastomer, were cross-linked (cured) by hexamethylenediamine carbamate or BPA. They have an OI in the 48 to 60 range and are rated V-0 by UL.

Peroxide-curable copolymers of VDF or TFE with perfluoro(methyl vinyl ether) ($CF_2{=}CFOCF_3$) were introduced by Du Pont, under the trade names Viton GLT and Kalrez in the 1960s. About 4540 t of these fluoroelastomers is produced annually worldwide. Properties of typical fluoroelastomers are shown in Table 15.5.

Copolymers of TFE and HFP are sold by Du Pont under the trade name Teflon FEP and by Daikin Kogyo under the trade name

TABLE 15.4 Effect of Composition on the T_g of CTFE-VDF Copolymers

Weight % VDF in copolymer	T_g, °F
0	113
15	68
40	18
57	4

TABLE 15.5 Properties of Typical Fluoroelastomers

Property, units*	PFA	FEP	Fluorosilicone
Melting point T_m, °F	590	486	—
Glass transition temp. T_g, °F	—	—	—
Processing temp., °F	500	500	—
Molding pressure, 10^3 lb/in^2	6	6	—
Mold shrinkage, 10^{-3} in/in	d15	15	—
Heat deflection temp. under flexural load of 264 lb/in^2, °F	166	—	—
Maximum resistance to continuous heat, °F	350	—	400
Coefficient of linear expansion, 10^{-6} in/(in · °F)	66	60	—
Compressive strength, 10^3 lb/in^2	—	—	—
Izod impact strength, ft · lb/in of notch	—	3	—
Tensile strength, 10^3 lb/in^2	4	3	1.5
Flexural strength, 10^3 lb/in^2	—	—	—
Percent elongation	300	300	200
Tensile modulus, 10^3 lb/in^2	—	1	2
Flexural modulus, 10^3 lb/in^2	120	95	—
Shore D hardness	64	55	52
Specific gravity	2.15	2.1	1.7
Percent water absorption	0.03	0.01	0.01
Dielectric constant	2.0	2.0	3.0
Dielectric strength, V/mil	500	500	500
Resistance to chemicals at 75°F:†			
Nonoxidizing acids (20% H_2SO_4)	S	S	Q
Oxidizing acids (10% HNO_3)	S	S	Q
Aqueous salt solutions (NaCl)	S	S	S
Polar solvents (C_2H_5OH)	S	S	S
Nonpolar solvents (C_6H_6)	S	S	S
Water	S	S	S
Aqueous alkaline solutions (NaOH)	S	S	S

*lb/(in^2 · 0.145) = KPa (kilopascals); ft · lb/(in of notch · 0.0187) = cm · N/cm of notch.
†S = satisfactory; Q = questionable; U = unsatisfactory.

Neoflon. Fluorinated copolymers of ethylene propylene (FEP) are also available commercially.

Other commercial polyfluorocarbons are copolymers of TFE (50%) and ethylene (50%) (Tefzel, Afton, Hostaflon ET), and copolymers of TFE and propylene (Aflas). Aflas has a T_g value of 2°C (36°F).

In spite of the excellent resistance of these elastomers to heat and solvents, they are not flexible at low temperatures. This deficiency was overcome in the 1960s by the synthesis of fluorosilicones ($\text{+}CH_3(CH_2CH_2CF_3)SiO\text{+}$; Silastic). Other commercial elastomers, with low T_g values [−51°C (− 60°F)] are the copolymers of TFE and trifluoronitrosomethane. However, these nitroso elastomers are not resistant to alkalies and decompose at temperatures above 163°C (325°F).

15.10 Applications of Polyfluorocarbons

Since many of these polyfluorocarbons can be used continuously at 260°C (500°F) and have low coefficients of friction, outstanding resistance to corrosives, and excellent electrical properties, they are used where no other available material of construction is applicable. These polymers are used as pipe liners, pipe fittings, pump fittings, and seals and valve liners, and in radar and electrical applications.

15.11 Glossary

aerosol A gaseous suspension of ultramicroscopic particles of liquid or solid.

Aflas A copolymer of TFE and propylene.

Afton A copolymer of TFE and ethylene.

ammonia NH_3

amorphous Noncrystalline.

bisphenol A $HOC_6H_4C(CH_3)_2C_6H_4OH$

coefficient of friction μ = force/load = F/W

copolymer A macromolecule having two repeating units.

covalent bond A bond formed by sharing electrons between two atoms.

FEP Fluorinated copolymer of ethylene propylene: $-\!\!\!\!+CF_2CF_2CF_2CF(CF_3)+\!\!\!\!-$

Fluon A PTFE.

Fluorel A VDF-HFP copolymer.

fluorine F.

Freon Volatile fluoroalkanes, such as dichlorodifluoromethane (CCl_2F_2).

Halon A TFE-HFP copolymer.

hexafluoropropylene $F_2C{=}CF(CF_3)$; HEP.

HFP Hexafluoropropylene.

Hostaflon ET A copolymer of TFE and ethylene.

Kalrez A copolymer of TFE and hexafluoromethyl vinyl ether.

Kel-F A PCTFE.

Kynar A PVDF.

Neoflon A copolymer of TFE and HFP.

NR Natural rubber.

oxygen index OI, the highest oxygen content in an oxygen-nitrogen mixture at which a candlelike flame does not burn.

pendant group A group attached to the principal polymer chain by a covalent bond.

perfluoromethyl vinyl ether $CF_2{=}CFOCF_3$

PET Polyethylene terephthalate.

PFA Perfluoroalkoxy polymer: $-(CF_2CF_2CF(OR)CF_2CF_2)-$

piezoelectric Transforming mechanical to electric energy.

Plunkett, R. The inventor of Teflon (PTFE).

polytetrafluoroethylene PTFE.

PTFE Polychlorotrifluoroethylene: $-(F_2CCFCl)-_n$.

PVDF Polyvinylidine fluoride: $-(CH_2CF_2)-_n$; Kynar.

PVF Polyvinyl fluoride: $-(CH_2CHF)-_n$.

Silastic A fluorosilicone.

sintering The formation of a coherent bonded mass by heating an aggregate, under pressure, below its melting point.

syndiotactic A polymer with pendant groups on one side of the polymer chain.

Tedlar A PVF.

Teflon A PTFE.

Tefzel A copolymer of TFE and ethylene.

tetrafluoroethylene $F_2C{=}CF_2$

TFE Tetrafluoroethylene.

T_g Glass transition temperature, i.e., the temperature at which a brittle plastic becomes flexible as the temperature is increased.

UL Underwriters' Laboratory.

uranium hexafluoride UF_6

vinylidine fluoride VDF: $H_2C{=}CF_2$.

Viton A A copolymer of HFP and VDF.

Viton B A terpolymer of VDF-HFP-TFE.

Viton GLT A copolymer of VDF and perfluoromethyl vinyl ether.

15.12 References

Chandrasekaran, S., and W. A. Milton: in "Abrasion and Wear," in J. I. Kroschwitz (ed.): *Encyclopedia of Polymer Science and Engineering,* vol. 3, Interscience Publishers, John Wiley & Sons, Inc., New York, 1985.

Lancaster, J. K.: in "Chlorotrifluoroethylene Polymers," in J. I. Kroschwitz (ed.): *Encyclopedia of Polymer Science and Engineering,* vol. 1, Interscience Publishers, John Wiley & Sons, Inc., New York, 1985.

Plunkett, R. J., C. A. Sperati, R. E. Putnam, J. E. Dohany, and H. Schroeder: in R. B. Seymour and G. S. Kirschenbaum (eds.): *High Performance Polymers: Their Origin and Development,* Elsevier Science Publishing Co., Inc., New York, 1986.

Rudner, M. A.: *Fluorocarbons,* Reinhold Publishing Corporation, New York, 1958.

Chapter

16

Polyimides and Ester and Amide Copolymers

16.1 Introduction

PIs, which contain both aromatic rings and heterocyclic rings in the

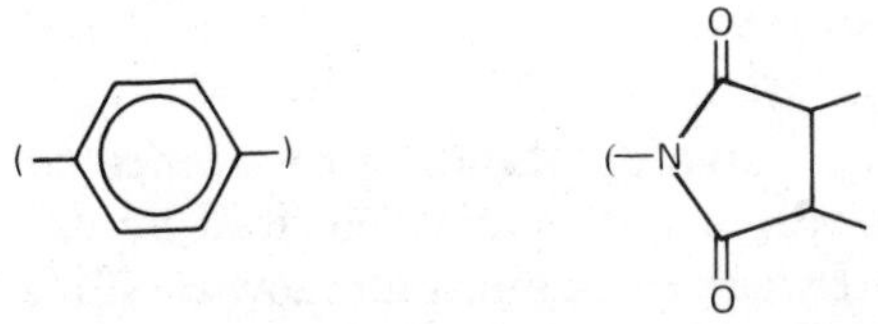

half ladder–ladder polymer chain, were synthesized by M. Bogart and R. Renshaw in 1908, but commercial development was delayed until there was a definite need for temperature-resistant resins in aerospace applications. H film (Kapton), which has a heat deflection temperature of 279°C (535°F), and Vespel have been available for a few decades from Du Pont.

Some PIs are thermoplastic, but most are produced by the imidization of a fusible polyamic acid. The latter is the reaction product of a primary amine and an anhydride or ester. The trade names Skybond, Pyralin, Thermid, Kerimid, and Kinel are used for PI resins. Ciba-Geigy has introduced a new bis-maleimide polymer which is being used in high-performance turbine engines.

Imidized PIs are commercially available from Ciba-Geigy and Upjohn. Considerable engineering data are also available on an experimental PI which is called Larc-TPI.

16.2 Processing of PI

Skybond, Pyralin, and NR-150 are available as PI prepolymers in which the rings are closed (imidized) during processing. This imidization stiffens the polymer chains and reduces flow in the mold.

Kerimid, Thermid, and PMR are available as low-molecular-weight prepolymers with acetylene end groups. Acetylene-terminated oligomers (Thermid) can be cured at 199°C (390°F) to produce strong void-free PIs that are stable at temperatures up to 371°C (700°F).

The imidized PIs, as supplied by Ciba-Geigy and Upjohn, are readily injection-molded. The more rigid PIs are processed by compression molding at 260°C (500°F) and heated subsequently at 343°C (650°F) or autoclave-molded at 260°C (500°F). Vacuum is applied to the autoclave-molding process to remove volatiles which could produce voids in the molded parts.

Large parts, such as transmission thrust bearings, have been processed by sintering techniques, i.e., pressing finely divided PI resin into the final shape, and then curing this billet at 360°C (680°F).

Laminates, fibers, and films (Kapton) are processed from solutions of PIs. In the lamination process, fiberglass, quartz, carbon, boron, or aramid fibers are impregnated with PI and compression-molded. Films of controlled thickness are produced on rollers.

16.3 Properties of PIs

PIs are tough polymers which have outstanding resistance to heat, gamma radiation, and flame. PIs are also characterized by excellent dielectric properties, excellent wear resistance, and low coefficients of expansion. However, they are soluble in highly polar solvents and are not stable to UV radiation. A tough PI, based on *N*-methylamine–imidized PMMA (Kamax), which is available from Rohm & Haas, lacks the thermal-resistance properties of the conventional PI products.

PI has a limiting oxygen index (LOI) of 30 to 38. The LOI for combustible substances is the oxygen-nitrogen content of air (21%), and the resistance to flame increases as the LOI increases. This is a vertical, i.e., candlelike, burning test, which corresponds with the UL 94 V-0 test in which PI is rated as V-1. However, this value varies depending on the type of PI being tested.

PI retains much of its good mechanical properties after long-time exposure at 204°C (400°F). Its half-life is 20,000 h in class H [180°C (356°F)]. Its wear resistance is excellent and similar to that of a type ASZ UN alloy. The coefficient of friction of a typical PI (Kinel 5505) decreases from 0.2 at 52°C (125°F) to 0.15 at 204°C (400°F).

TABLE 16.1 Properties of Typical PIs

Property, units*	PI	PI + 40% graphite
Melting point T_m, °F	—	—
Glass transition temp. T_g, °F	635	690
Processing temp., °F	650	690
Molding pressure, 10^3 lb/in^2	5	5
Mold shrinkage, 10^{-3} in/in	3	2
Heat deflection temp. under flexural load of 264 lb/in^2, °F	600	680
Maximum resistance to continuous heat, °F	600	625
Coefficient of linear expansion, 10^{-6} in/(in · °F)	30	20
Compressive strength, 10^3 lb/in^2	30	18
Izod impact strength, ft · lb/in of notch	1.5	0.7
Tensile strength, 10^3 lb/in^2	14	7.5
Flexural strength, 10^3 lb/in^2	24	14
Percent elongation	8	3
Tensile modulus, 10^3 lb/in^2	300	—
Flexural modulus, 10^3 lb/in^2	500	700
Rockwell hardness	E70	E27
Specific gravity	1.4	1.6
Percent water absorption	0.25	0.15
Dielectric constant	5	—
Dielectric strength, V/mil	560	—
Resistance to chemicals at 75°F:†		
Nonoxidizing acids (20% H_2SO_4)	Q	Q
Oxidizing acids (10% HNO_3)	Q	Q
Aqueous salt solutions (NaCl)	S	S
Polar solvents (C_2H_5OH)	S	S
Nonpolar solvents (C_6H_6)	S	S
Water	S	S
Aqueous alkaline solutions (NaOH)	U	U

*lb/(in^2 · 0.145) = KPa (kilopascals); ft · lb/(in of notch · 0.0187) = cm · N/cm of notch.
†S = satisfactory; Q = questionable; U = unsatisfactory.

PI retains its electrical properties over a wide range of frequencies. Thus, the dielectric constant is 4.84 at 24°C (75°F) at a frequency of 50 Hz and 4.74 at a frequency of 1×10^6 Hz. There is a slight increase at 1×10^6 Hz to 4.82 at 204°C (400°F). Its physical properties are essentially unchanged by exposure to a radiation of 10^{10} rads. Because of the many formulations commercially available, it is essential to obtain recommendations from the supplier before designing a PI part of a high-performance application.

The properties of typical PIs are shown in Table 16.1.

16.4 PAI

The processibility of PI resins has been improved by copolymerization to produce PAIs (Torlon), transparent PI sulfones (Duratherm), PEIs

(Ultem), and PI sulfones (PISO). The use of the PAI in the polimotor has reduced the weight by 50 percent. PAI is injection-moldable and has a heat deflection temperature of 191°C (375°F). PISO has a heat deflection temperature of 210°C (410°F).

As shown by the repeating unit in Fig. 16.1, PAI includes an amide group ⟶CONH⟶ in addition to the imide group.

This injection-moldable PAI (Torlon) can be used continuously at 221°C (430°F) in contrast to the sintered PI (Vespel), which can be used continuously at 260°C (500°F). PAI wire enamels have received an Institute of Electrical and Electronics Engineers (IEEE) 57 thermal rating of 221°C (430°F) for 20,000 h. Amoco, which is the supplier of Torlon, claims that PAI has the highest strength of any commercial unreinforced plastic. PAI can be postcured at 260°C (500°F).

As was the case for PI, there are many different grades of PAI, of which the 9000 series represents the most recently developed products. Other series are high-strength and wear-resistant grades. The inherent lubricity of high-strength grades, such as 42032, is enhanced by the addition 0.5 percent PTFE. This grade also contains 3 percent titanium dioxide (TiO_2) pigment. PAI is also available as an injection-moldable, glass-reinforced (Torlon 5030) and graphite fiber–reinforced (Torlon 7130) creep-resistant high-performance plastic.

In addition to its excellent performance at elevated temperatures, PAI has excellent properties under cryogenic conditions [− 160°C (− 320°F)]. For example, Torlon 4203L has a tensile strength of 31,500

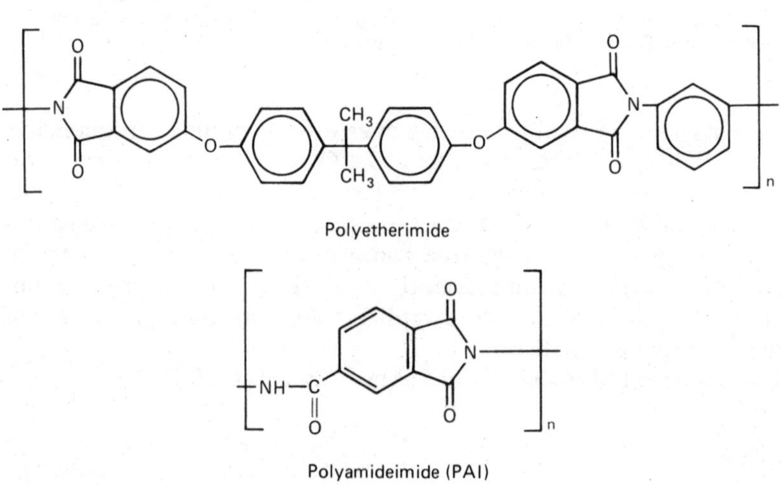

Polyetherimide

Polyamideimide (PAI)

Figure 16.1 Repeating unit of a PEI and a PAI.

lb/in^2, an elongation of 6 percent, a flexural strength of 41,000 lb/in^2, and a flexural modulus of 1,140,000 lb/in^2. Other PAIs, such as Torlon 7130, have higher flexural modulus (3,000,000 lb/in^2), and this modulus is retained at 149°C (300°F) but decreases to 2,500,000 lb/in^2 at 232°C (450°F).

As shown in Fig. 16.2, Torlon 4203L has a 15 percent elongation at a tensile stress of about 25,000 lb/in^2, but Torlon 7130, 9040, and 5030

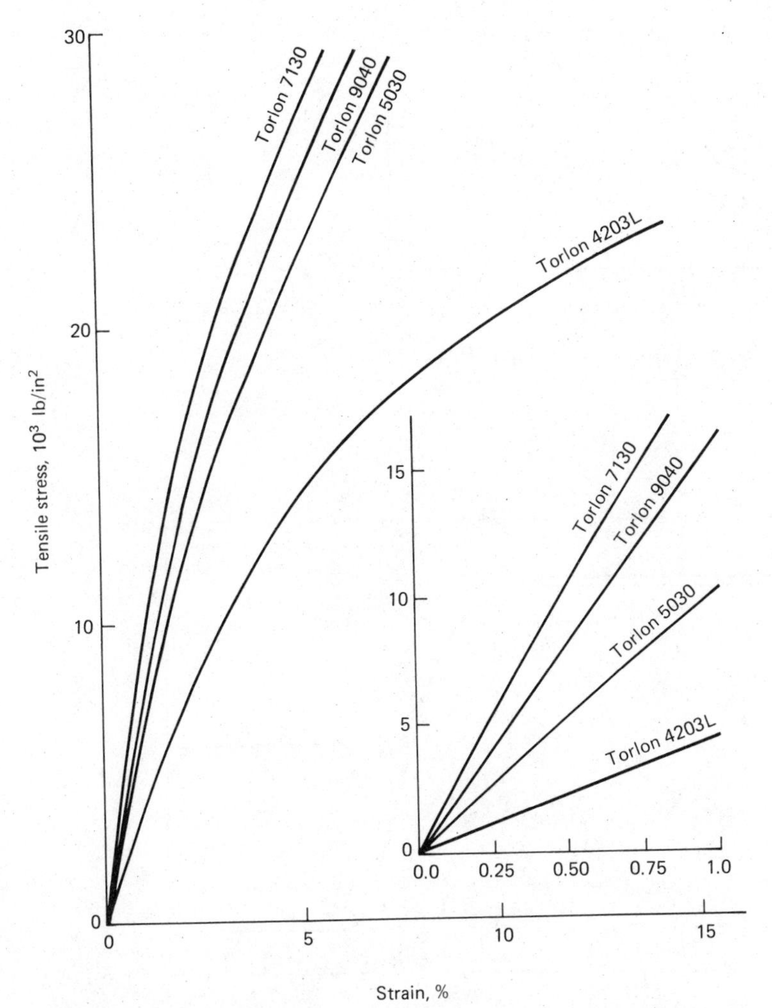

Figure 16.2 Percent elongation and tensile stress for Torlon 4203L, 7130, 9040, and 5030.

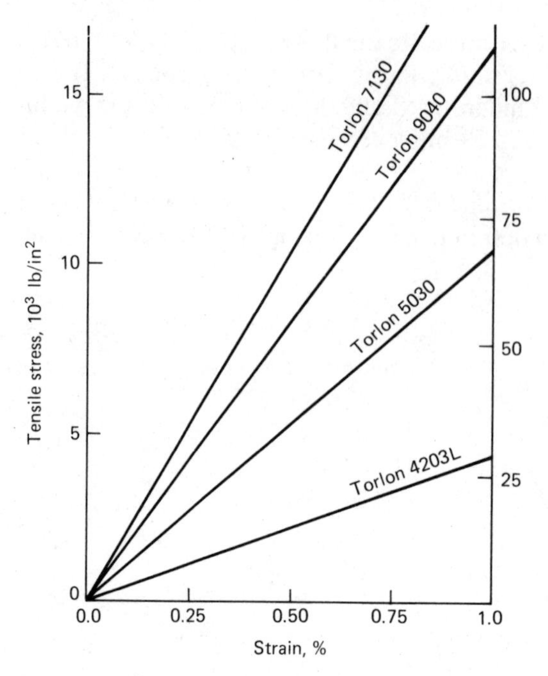

Figure 16.3 Stress-strain relationship of PAIs.

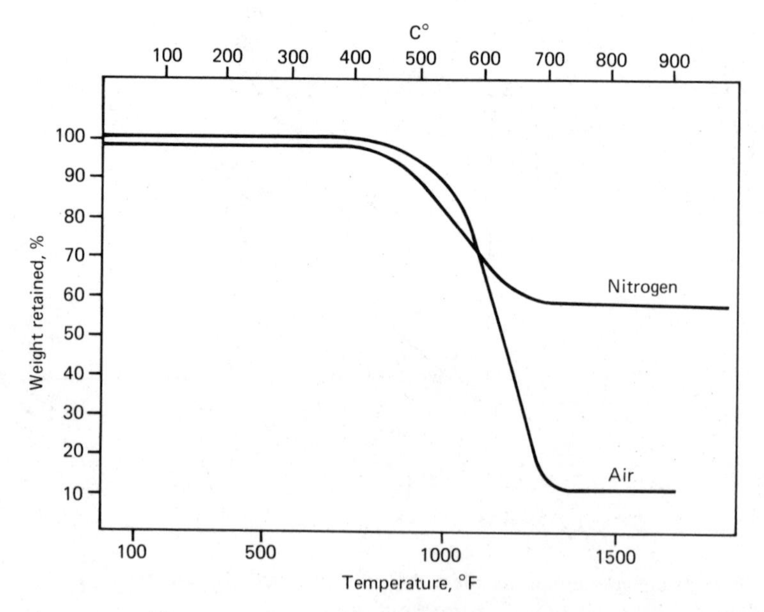

Figure 16.4 Thermogravimetric analysis of Torlon 4203L.

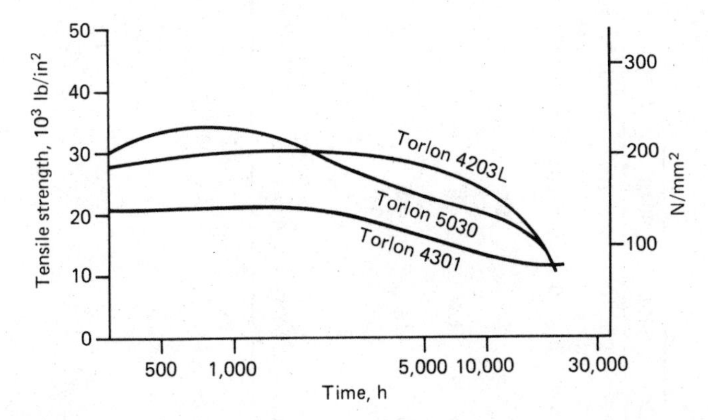

Figure 16.5 Tensile strength of PAI after thermal aging at 250°C (482°F).

have elongations of less than 5 percent at a tensile stress of about 30,000 lb/in^2. As shown in Fig. 16.3, the stress-strain relations of these PAIs are essentially linear (hookean) at 135°C (275°F).

Since PAI molded parts are used on a volume, rather than a weight, basis, specific strength, i.e., strength/density, is important to the designer.

As shown in Fig. 16.4, there is essentially no weight loss when Torlon 4203L is heated at about 482°C (900°F). The UL thermal index predicts a useful life of 100,000 h for commercial PAIs at temperatures of about 204°C (400°F). This retention of tensile strength at 250°C (482°F) is illustrated in Figs. 16.4 and 16.5. Also as shown by the data in Table 16.2, the coefficients of expansion [10^{-6} in/

TABLE 16.2 Coefficients of Expansion [10^{-6} in/(in · °F)] of Selected Materials

Torlon 7130	5
Torlon 9040, 7330	7
Torlon 5030	9
Torlon 5030M, 5430	10
Torlon 4275, 4301	14
Torlon 4347	15
Torlon 4203L	17
Inconel X, annealed	6.7
Carbon steel	6.7
Titanium	7
Copper	9
Stainless steel	9.6
Bronze	10.2
Aluminum alloy	12.7
Torlon 2017, 7075	14.4

TABLE 16.3 Electrical Properties of Insulating Grades of Torlon Resin

	Torlon grade									
	4203L	4301*	4275*	4347*	5030	5030M	5430	9040	7130	7233
Volume resistivity (ASTM D257), Ω · in	8×10^{16}	3×10^{15}	3×10^{15}	3×10^{15}	6×10^{16}	2×10^{16}	2×10^{16}	2×10^{16}	3×10^{6}*	8×10^{5}*
Surface resistivity (ASTM D257), Ω · in	5×10^{18}	8×10^{17}	4×10^{17}	1×10^{18}	1×10^{18}	8×10^{16}	3×10^{17}	9×10^{17}	5×10^{7}*	9×10^{6}*
Dielectric strength, 0.040 in (ASTM D149), V/mil	580	—	—	—	840	730	650	490		
Dielectric constant (ASTM D150)	4.2	6.0	7.3	6.8	4.4	4.0	4.2	4.3		
At 1000 Hz	4.2	6.0	7.3	6.8	4.4	4.0	4.2	4.3		
At 1,000,000 Hz	3.9	5.4	6.6	6.0	6.5	4.3	4.0	4.6		
Dissipation factor (ASTM D150)										
At 1000 Hz	0.026	0.037	0.059	0.037	0.022	0.040	0.020	0.040		
At 1,000,000 Hz	0.031	0.042	0.063	0.071	0.023	0.030	0.036	0.044		

*Contain graphite powder. By these tests, they behave as insulators, but they may behave in a more conductive manner at high voltage or high frequency.

†Molded specimen—plaque 3.0 × 4.5 × 0.190 in (7.6 × 11 × 0.48 cm).

(in · °F)] of commercial PAIs are in the same order of magnitude as those of many metals.

The OI varies from 45 percent for Torlon 4203L to 52 percent for Torlon 7130. Less than 10 ppm carbon monoxide (CO) is emitted from smoldering PAI, and this emission from flaming PAI is less than 120 ppm. The flash ignition temperature and self-emission temperature for a typical PAI (Torlon 4203L) are 570 and 620°C (1058 and 1148°F), respectively.

Small dimensional changes occur when molded PAI is exposed to high humidity over a relatively long period of time (100 + days). This effect is reversible, and the original dimension can be restored by heating at 149°C (300°F) for 16 h. Care must be used when heating PAIs under high humidity conditions to prevent thermal shock. PAI is resistant to weathering and gamma radiation. The electrical properties of PAIs are shown in Table 16.3.

The wear resistance of wear-resistant grades of PAI is similar to that of PI. A postcure is essential for maximum wear resistance.

Most of the information on PAI was generously supplied by Amoco, and this firm should be consulted for more specific information on PAI. The properties of typical PAIs are shown in Table 16.4.

16.5 PEI

In 1967, Gorvin used sodium salts of phenols to displace the nitro groups in dinitrobenzophenones, and in 1969, Rudman and other General Electric chemists obtained polyether ketones by comparable reactions. Subsequently, J. Wirth applied these concepts for the production of a PEI from the reaction of BPA, 4-nitrophthalamide, and *m*-phenylenediamine.

This exceptionally strong, amorphous, amber-colored transparent, high-performance polymer is sold by General Electric under the trade name Ultem. In spite of its high heat deflection temperature [(217°C (423°F)], PEI is readily injection-molded and extruded. According to General Electric, the OI of Ultem (47%) is higher than that of any other engineering polymer. The flash ignition temperature of PEI is 521°C (970°F), and the UL 94 rating is V-0. The properties of PEI are shown in Table 16.5.

16.6 Applications of Imide Copolymers

PAI and PEI are used as computer gears and discs, surgical and diagnostic instruments, aircraft and automotive components, microwave

TABLE 16.4 Properties of Typical PAIs

Property, units*	PAI	PAI + 30% fiberglass	PAI + 30% graphite filler
Melting point T_m, °F	—	—	—
Glass transition temp. T_g, °F	527	527	527
Processing temp., °F	650	650	650
Molding pressure, 10^3 lb/in^2	35	35	35
Mold shrinkage, 10^{-3} in/in	7	3	1
Heat deflection temp. under flexural load of 264 lb/in^2, °F	532	539	540
Maximum resistance to continuous heat, °F	500	500	500
Coefficient of linear expansion, 10^{-6} in/(in · °F)	20	15	15
Compressive strength, 10^3 lb/in^2	32	38	37
Izod impact strength, ft · lb/in of notch	2.7	1.5	1.0
Tensile strength, 10^3 lb/in^2	22	32	36
Flexural strength, 10^3 lb/in^2	30	48	50
Percent elongation	7.6	2.3	1.2
Tensile modulus, 10^3 lb/in^2	650	2110	3570
Flexural modulus, 10^3 lb/in^2	730	1700	2780
Rockwell hardness	E86	E96	E94
Specific gravity	1.4	1.6	1.5
Percent water absorption	0.3	0.25	0.25
Dielectric constant	4.2	4.2	6.0
Dielectric strength, V/mil	580	840	—
Resistance to chemicals at 75°F:†			
Nonoxidizing acids (20% H_2SO_4)	S	S	S
Oxidizing acids (10% HNO_3)	S	S	S
Aqueous salt solutions (NaCl)	S	S	S
Polar solvents (C_2H_5OH)	S	S	S
Nonpolar solvents (C_6H_6)	S	S	S
Water	S	S	S
Aqueous alkaline solutions (NaOH)	Q	Q	Q

*lb/(in^2 · 0.145) = KPa (kilopascals); ft · lb/(in of notch · 0.0187) = cm · N/cm of notch.
†S = satisfactory; Q = questionable; U = unsatisfactory.

oven hardware, pressure vessels, sleeve bearings, electric switches, plenum connectors, and circuit boards.

16.7 Glossary

acetylene HC≡CH

amide group —CONH—

amorphous Noncrystalline.

aramid Aromatic nylon (polyamide).

TABLE 16.5 Properties of Typical PEIs

Property, units*	Ultem 1000 (unfilled)	Ultem 2100 (10% fiber-glass)	Ultem 2200 (20% fiber-glass)	Ultem 2400 (40% fiber-glass)
Melting point T_m, °F	—	—	—	—
Glass transition temp. T_g, °F	421	421	421	421
Processing temp., °F	660	660	660	660
Molding pressure, 10^3 lb/in^2	15	15	15	15
Mold shrinkage, 10^{-3} in/in	5	3.5	2.5	2.0
Heat deflection temp. under flexural load of 264 lb/in^2, °F	390	405	408	415
Maximum resistance to continuous heat, °F	350	375	375	380
Coefficient of linear expansion, 10^{-6} in/(in · °F)	30	10	14	8
Compressive strength, 10^3 lb/in^2	20	22	24	32
Izod impact strength, ft · lb/in of notch	1.0	1.1	1.6	2.1
Tensile strength, 10^3 lb/in^2	14	16	20	27
Flexural strength, 10^3 lb/in^2	22	28	30	36
Percent elongation	60	5	3	2.5
Tensile modulus, 10^3 lb/in^2	430	650	1000	1700
Flexural modulus, 10^3 lb/in^2	480	650	900	1700
Rockwell hardness	M109	M114	M114	M114
Specific gravity	1.27	1.34	1.42	1.61
Percent water absorption	0.25	0.20	0.15	0.13
Dielectric constant	3.1	3.5	3.5	3.7
Dielectric strength, V/mil	480	700	670	610
Resistance to chemicals at 75°F:†				
Nonoxidizing acids (20% H_2SO_4)	S	S	S	S
Oxidizing acids (10% HNO_3)	U	U	U	U
Aqueous salt solutions (NaCl)	S	S	S	S
Polar solvents (C_2H_5OH)	S	S	S	S
Nonpolar solvents (C_6H_6)	S	S	S	S
Water	S	S	S	S
Aqueous alkaline solutions (NaOH)	Q	Q	Q	Q

*lb/(in^2 · 0.145) = KPa (kilopascals); ft · lb/(in of notch · 0.0187) = cm · N/cm of notch.
†S = satisfactory; Q = questionable; U = unsatisfactory.

creep Long-term flow.

cryogenic Low-temperature.

dielectric constant A measure of the ability of molecules to become polarized in an electric field, i.e., the ratio of the capacitance of a condenser filled with the test materials to the capacitance of an evacuated capacitor.

dielectric strength The maximum voltage before failure (dielectric breakdown).

dissipation factor A measure of the energy dissipation (dielectric loss) to heat.

Duratherm A PISO.

fiberglass Glass fiber.

H film A PI.

half-life The time to obtain 50 percent of the original property.

heterocyclic ring A ring containing carbon and nitrogen, oxygen, or sulfur atoms.

hookean Following Hooke's law: stress is proportional to strain.

Hz Hertz (measure of frequency).

Kamax A PI.

Kapton A PI.

Kerimid A PI.

Kinel A PI.

imidization Cyclization to form a heterocyclic ring.

ladder polymer A double-stranded polymer.

laminate A sheet consisting of several bonded layers.

Larc-TPI A PI.

limiting oxygen index The highest percentage of oxygen in an oxygen-nitrogen mixture which does not support combustion of a test material.

LOI Limiting oxygen index.

NR-150 A PI.

oligomer A low-molecular-weight polymer.

PAI Polyamide imide.

PI Polyimide.

PISO Polyimide sulfone.

PMR A PI.

polimotor An experimental gasoline combustion engine.

Pyralin A PI.

sintering The compression of powdery material to form a billet.

Skybond A PI.

specific strength The strength/density ratio.

surface resistivity The resistance to electric current along a 1 cm^2 surface of the test material.

Thermid A PI.

Torlon A PAI.

UL Underwriters' Laboratory.

Vespel A PI.

volume resistivity The electric resistance (ohms) between faces of a cube times the area of the faces.

16.8 References

Deanin, R. D.: *New Industrial Polymers,* Marcel Dekker, Inc., New York, 1975.
Fitzpatrick J. E.: "Engineering Polymers," in J. N. Epel, J. M. Margolis, S. Newman, and R. B. Seymour (eds.): *Engineered Materials Handbook*, vol. 2, ASM International, Metals Park, Ohio, 1988.
King, J. J., and B. H. Lee: in "Polymides," R. B. Seymour and G. S. Kirshenbaum (eds.): *High Performance Polymers: Their Origin and Development,* Elsevier Science Publishing Co., Inc., New York, 1986.
Margolis, J. M.: *Engineering Thermoplastics,* Marcel Dekker, Inc., New York, 1985.
Mittal, K. L.: *Polyimides: Synthesis, Characterization and Applications,* Plenum Press, Plenum Publishing Corporation, New York, 1984.
Seymour, R. B.: *Polymers for Engineering Applications,* ASM International, Metals Park, Ohio, 1986.
Wirth, J. G.: "Polyetherimides," in R. B. Seymour and G. S. Kirshenbaum (eds.): *High Performance Polymers: Their Origin and Development,* Elsevier Science Publishing Co., Inc., New York, 1986.

Chapter

17

Polyether Ketones, Polysulfones, and Polyphenylene Sulfide

17.1 Polyether Ketones

Goodman of ICI and Bonner of Du Pont developed polyether ketone (PEK) technology in the early 1960s. Raychem introduced a PEK (Stilan) commercially in the early 1970s. Stilan was acquired by BASF. Rose of ICI produced a PEEK (Victrex) in the early 1980s. PEK and PEEK have heat deflection temperatures of 166 and 149°C (330 and 300°F), respectively. BASF, Amoco, and Hoechst Celanese also produce aromatic ketone polymers under the trade names Ultrapek, Kadel, and Hostatec.

The carbonyl (C═O) group in both PEK and PEEK acts as a stiffening group, and the ether group (O) provides a small degree of flexibility in PEEK. Both PEK and PEEK are readily injection-moldable and extrudable using conventional machines.

PEEK has a high degree of oxidative and radiation stability and has been used for years at a continuous temperature of over 199°C (390°F). It meets UL 94 V-0 requirements and has a UL index of 210°C (410°F). It has an LOI rating of 34 and has low smoke emission. PEEK is resistant to nonoxidizing acids, such as hydrochloric acid, alkalies, salts, and solvents. It is attacked by concentrated sulfuric acid, nitric acid, and chlorine.

The structural formulas for these high-performance polymers are shown in Fig. 17.1.

PEEK has excellent resistance to cyclic stress, and its tensile strength decreases less than 10 percent after 10^7 cycles at 23°C (73°F). The flexural moduli of PEK and PEEK remain unchanged up to a

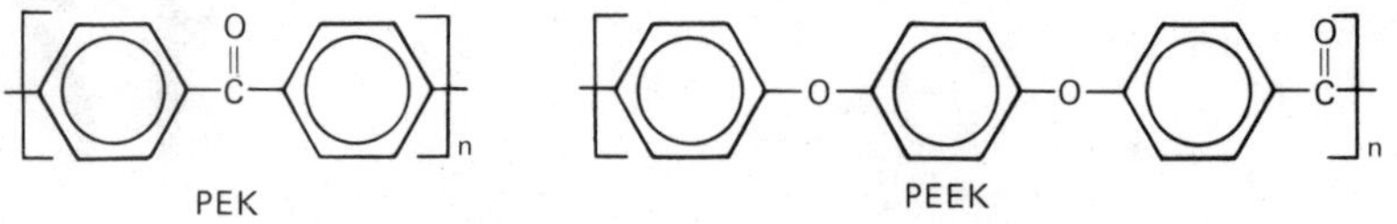

PEK PEEK

Figure 17.1 Structural formula for high-performance polymers.

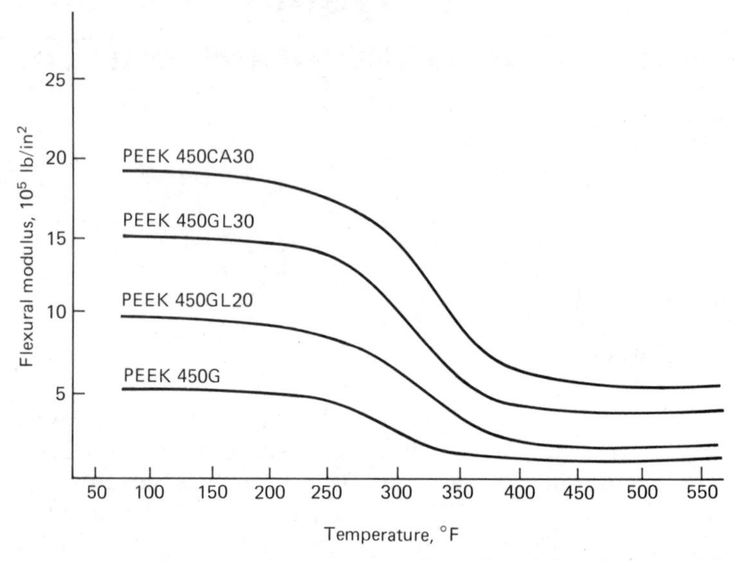

Figure 17.2 Effect of temperature on the flexural modulus of commercial PEEKs.

temperature of 121°C (250°F). The effect of temperature on the flexural modulus of commercial Victrex PEEK is shown in Fig. 17.2.

Also, as shown by the plots in Fig. 17.3, there is very little change in tensile strength when PEEK is immersed in water at 100°C (212°F) for as long as 200 days.

Since they are crystalline polymers, the strength and thermal resistance of PEK and PEEK are increased dramatically by the incorporation of reinforcing agents. The resistance of PEK and PEEK to weathering and gamma radiation is excellent. The properties of typical PEK and PEEK plastics are shown in Table 17.1.

17.2 Applications of PEK and PEEK

The initial application of PEEK was as extruded insulation for wire and cable; this was followed by injection-molded components of PEK

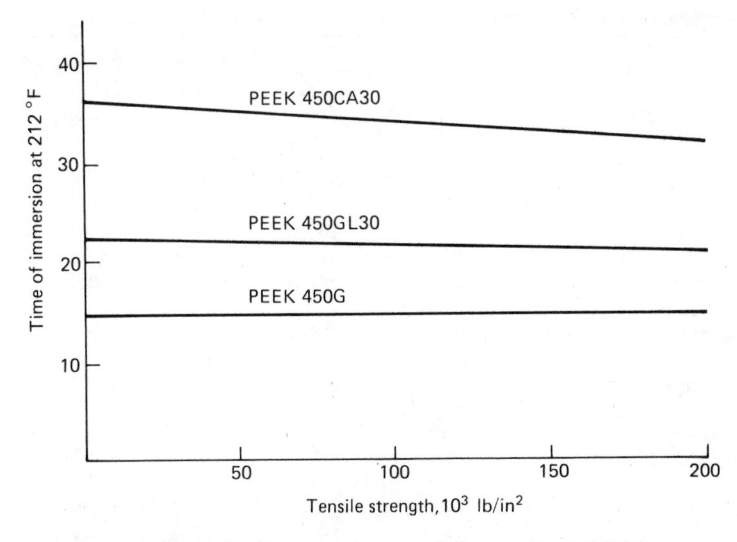

Figure 17.3 Effect of boiling water on tensile strength of PEEK.

and PEEK for jet engines, aircraft, and automobiles, liners for bearings, extruded film (Stabar), yarn (Zyex), and carbon fiber prepreg (APC).

17.3 Polysulfones

Unlike many general-purpose polymers, which are synthesized and commercialized without much basic knowledge of polymer science, several condensation polymers, including nylon, aramids, polyketones, and polysulfones, were developed by polymer chemists who were cognizant of the relations between structure and polymer properties.

Polysulfone (Udel) was produced by R. N. Johnson at Union Carbide in 1966, and other sulfones were introduced in 1967 by 3M (Astrel), in 1972 by Union Carbide (Radel), and in 1971 by ICI (Victrex PES). These products, which were developed independently, differ primarily in the flexibility of their polymer chains.

Udel, which was produced by the condensation of BPA and 4,4′-dichlorodiphenylsulfone, is an amber transparent polymer with a heat deflection temperature of 174°C (345°F). Victrex PES, which is a polyether sulfone (PES), is also a transparent yellow polymer with a heat deflection temperature of 203°C (397°F) and a UL rating of 180°C (356°F). The heat deflection temperature of 40 percent fiberglass-filled PES is 180°C (356°F). Udel is now being produced by Amoco. The annual production of polysulfones is 4000 t, and PES accounts for much of this production.

TABLE 17.1 Properties of Typical PEK and PEEKs

Property, units*	PEK	PEEK 450G	PEK + 30% fiberglass	PEEK + 30% fiberglass
Melting point T_m, °F	690	635	—	—
Glass transition temp. T_g, °F	330	290	290	290
Processing temp., °F	750	700	750	800
Molding pressure, 10^3 lb/in^2	15	15	15	15
Mold shrinkage, 10^{-3} in/in	10	10	7.5	7.5
Heat deflection temp. under flexural load of 264 lb/in^2, °F	330	300	600	600
Maximum resistance to continuous heat, °F	300	275	500	500
Coefficient of linear expansion, 10^{-6} in/(in · °F)	20	20	12	12
Compressive strength, 10^3 lb/in^2	10	18	22	30
Izod impact strength, ft · lb/in of notch	1.5	1.5	2.3	1.7
Tensile strength, 10^3 lb/in^2	16	13	25	32
Flexural strength, 10^3 lb/in^2	16	16	36	44
Percent elongation	75	75	2.5	2.5
Tensile modulus, 10^3 lb/in^2	400	400	1400	2500
Flexural modulus, 10^3 lb/in^2	435	435	1400	1800
Rockwell hardness	R126	R126	—	—
Specific gravity	1.3	1.3	1.5	1.4
Percent water absorption	0.5	0.5	0.1	0.05
Dielectric constant	3.2	3.2	3.5	—
Dielectric strength, V/mil	480	480	480	—
Resistance to chemicals at 75°F:†				
Nonoxidizing acids (20% H_2SO_4)	S	S	S	S
Oxidizing acids (10% HNO_3)	S	S	S	S
Aqueous salt solutions (NaCl)	S	S	S	S
Polar solvents (C_2H_5OH)	S	S	S	S
Nonpolar solvents (C_6H_6)	S	S	S	S
Water	S	S	S	S
Aqueous alkaline solutions (NaOH)	S	S	S	S

*lb/(in^2 · 0.145) = KPa (kilopascals); ft · lb/(in of notch · 0.0187) = cm · N/cm of notch.
†S = satisfactory; Q = questionable; U = unsatisfactory.

There are several different structures for these amorphous yellow transparent aromatic polysulfones, but as shown by the formula in Fig. 17.4, they all contain sulfone (SO_2) stiffening groups, and most contain an ether flexibilizing group.

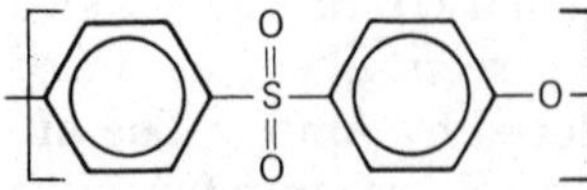

Figure 17.4 Typical structure of a polysulfone.

PES is resistant to attack by nonoxidizing acids, alkalies, salts, and aliphatic hydrocarbon solvents, such as gasoline, is dissolved in chlorinated hydrocarbons (carbon tetrachloride), is attacked by concentrated sulfuric acid, and is swollen in aromatic hydrocarbon solvents (benzene), esters (ethyl acetate), and ketones (acetone).

Like all other aromatic polymers, PES has limited resistance to photooxidation, but carbon black–filled stabilized PES has good weatherability. PES has been used for many years in hot water service and at temperatures up to 141°C (285°F). As is true for most polymers, the mechanical properties, such as impact resistance, decrease as the temperature is increased. PES has poor resistance to stress cracking in hot water and in corrosive environments. The properties of typical polysulfones are shown in Table 17.2.

The electrical properties of PES are essentially unchanged with temperature. Thus, the dielectric constant of PES is 3.37 at 20°C (68°F) and increases linearly to 3.44 at 127°C (260°F). Likewise, the loss tangent changes from 0.013 at 20°C (68°F) to 0.017 at 127°C (260°F). The electrical properties of PES are shown in Table 17.3.

17.4 Applications of Polysulfones

Because of their excellent UL thermal index of 160°C (320°F), creep resistance, transparency, and mechanical and electrical properties, polysulfones are used as printed circuit boards, integrated circuit carriers, coil bobbins, TV and stereo components, under-the-hood and aircraft components, microwave cookware, cooking appliances, pacemakers, refrigerators, filtration membranes, and pipes. Information on the most desirable formulations for specific applications should be obtained from suppliers.

17.5 Polyphenylene Sulfide

Phenylene sulfide oligomers were produced in 1880 by Friedel and Crafts, who reacted benzene and sulfur in the presence of aluminum chloride. In 1948, MaCallum produced PPS by the reaction of sulfur, dichlorobenzene, and sodium carbonate at 299°C (570°F). After purchasing the MaCallum patents, Dow investigated the Lenz self-condensation of copper-*p*-bromothiophenoxide but did not produce PPS commercially. Commercial PPS, under the trade name Ryton, was produced by Edmunds and Hill of Phillips Petroleum in 1967 by the condensation of sodium sulfide and dichlorobenzene.

PPS is a dark-colored, crystalline, temperature- and chemical-resistant polymer. Its heat deflection temperature (DTUL) of 135°C

TABLE 17.2 Properties of Typical Polysulfones

Property, units*	Polysulfone	PES	Polysulfone + 30% fiberglass	PES + 20% fiberglass
Melting point T_m, °F	—	—	—	—
Glass transition temp. T_g, °F	375	435	375	428
Processing temp., °F	650	650	650	650
Molding pressure, 10^3 lb/in^2	15	15	15	15
Mold shrinkage, 10^{-3} in/in	7	6	1.5	3
Heat deflection temp. under flexural load of 264 lb/in^2, °F	345	395	350	415
Maximum resistance to continuous heat, °F	320	350	320	380
Coefficient of linear expansion, 10^{-6} in/(in · °F)	30	30	15	15
Compressive strength, 10^3 lb/in^2	40	14	19	22
Izod impact strength, ft · lb/in of notch	1.2	1.4	1.1	1.5
Tensile strength, 10^3 lb/in^2	15	12	14	18
Flexural strength, 10^3 lb/in^2	15	18	20	25
Percent elongation	75	45	2	2
Tensile modulus, 10^3 lb/in^2	360	350	1350	860
Flexural modulus, 10^3 lb/in^2	390	360	1050	875
Rockwell hardness	R126	R126	—	—
Specific gravity	1.3	1.3	1.5	1.4
Percent water absorption	0.5	0.5	0.1	0.05
Dielectric constant	3.2	3.2	3.5	—
Dielectric strength, V/mil	480	480	480	—
Resistance to chemicals at 75°F:†				
Nonoxidizing acids (20% H_2SO_4)	S	S	S	S
Oxidizing acids (10% HNO_3)	U	U	U	U
Aqueous salt solutions (NaCl)	S	S	S	S
Polar solvents (C_2H_5OH)	S	S	S	S
Nonpolar solvents (C_6H_6)	U	U	U	U
Water	S	S	S	S
Aqueous alkaline solutions (NaOH)	S	S	Q	Q

*lb/(in^2 · 0.145) = KPa (kilopascals); ft · lb/(in of notch · 0.0187) = cm · N/cm of notch.
†S = satisfactory; Q = questionable; U = unsatisfactory.

(275°F) is increased by heating the molded articles and by reinforcing with 40 percent fiberglass. The fiberglass-reinforced PPS has a UL index of over 199°C (390°F).

Pultruded-graphite-reinforced PPS, with a melting point of 285°C (545°F) and an Izod impact resistance of 39 ft · lb/in of notch, is available. Hoechst Celanese is using Japanese know-how to produce PPS under the trade name Fortron. Ciba-Geigy and Phillips Petroleum have teamed up to produce PPS in Europe, and Toray and Phillips Petroleum have sponsored a joint venture with Kureha Chemical in Japan, which is the second largest market for PPS.

TABLE 17.3 Electrical Properties of PES

Dielectric constant	
At 60 Hz	3.5
At 1,000,000 Hz	3.5
Loss tangent	
At 60 Hz	0.001
At 1,000,000 Hz	0.0035
Volume resistivity	10^{16} Ω/cm
Tracking resistance	150 V
Arc resistance	50
High-voltage arc ignition	300 s

PPS can be ignited to produce a yellow-orange flame, but it does not support prolonged combustion. It is not soluble in any known solvent at temperatures below 204°C (400°F). Because of its simple repetitive geometry on its polymer chain, injection-molded PPS parts crystallize rapidly if the mold temperature is above 121°C (250°F). Amorphous moldings can be obtained if the melt is quenched rapidly in a mold at a temperature below 93°C (200°F).

Amorphous PPS moldings have a T_g value of 88°C (190°F) and are much more creep-resistant than crystalline PPS. Of course, amorphous PPS moldings crystallize if they are heated above the crystallization temperature [127°C (260°F)]. The creep modulus at any time t may be calculated from the following formula, in which the pressure P on a bar with a span L is inversely proportional to the moment of inertia of the beam, I, multiplied by the extent of the deformation Δ.

$$\text{Creep at time } t = \frac{PL^3}{48I\Delta}$$

PPS is resistant to neutron and gamma radiation. Hence in nuclear installations, its flexural strength and modulus are essentially unchanged when it is exposed to gamma radiation of 5×10^9 rad and neutron radiation of 1×10^9 rad. The properties of PPS are shown in Table 17.4.

The producers of Ryton have developed considerable information on the effect of temperature on properties of PPS and should be consulted when additional design information is needed.

As shown in Fig. 17.5, the stress-strain plot for Ryton R-4 is essentially hookean at 23°C (73°F), i.e., the stress is proportional to the strain (elongation).

However, as shown in Fig. 17.6, the tensile strength of both amorphous (low crystalline) and crystalline PPS (Ryton-R-107006A) decreases as the temperature is increased.

TABLE 17.4 Properties of Typical PPS Moldings

Property, units*	PPS	PPS + 40% fiberglass	PPS + 40% carbon fiber
Melting point T_m, °F	550	550	500
Glass transition temp. T_g, °F	190	190	190
Processing temp., °F	625	625	625
Molding pressure, 10^3 lb/in^2	10	10	10
Mold shrinkage, 10^{-3} in/in	7	3	0.5
Heat deflection temp. under flexural load of 264 lb/in^2, °F	275	490	475
Maximum resistance to continuous heat, °F	250	450	475
Coefficient of linear expansion, 10^{-6} in/(in · °F)	30	15	20
Compressive strength, 10^3 lb/in^2	16	24	26
Izod impact strength, ft · lb/in of notch	0.5	1.2	1.5
Tensile strength, 10^3 lb/in^2	9.5	20	29
Flexural strength, 10^3 lb/in^2	14	26	40
Percent elongation	1.5	1.5	1.5
Tensile modulus, 10^3 lb/in^2	480	1100	4400
Flexural modulus, 10^3 lb/in^2	550	1750	3900
Rockwell hardness	R123	R123	R123
Specific gravity	1.3	1.65	1.5
Percent water absorption	0.02	0.03	0.03
Dielectric constant	3.8	4.6	—
Dielectric strength, V/mil	380	400	—
Resistance to chemicals at 75°F:†			
Nonoxidizing acids (20% H_2SO_4)	S	S	S
Oxidizing acids (10% HNO_3)	S	S	S
Aqueous salt solutions (NaCl)	S	S	S
Polar solvents (C_2H_5OH)	S	S	S
Nonpolar solvents (C_6H_6)	S	S	S
Water	S	S	S
Aqueous alkaline solutions (NaOH)	S	Q	S

*lb/(in^2 · 0.145) = KPa (kilopascals); ft · lb/(in of notch · 0.0187) = cm · N/cm of notch.
†S = satisfactory; Q = questionable; U = unsatisfactory.

The effects of temperature on the flexural strength and flexural modulus of Ryton R-4 are shown in Figs. 17.7 and 17.8.

The notched and unnotched Izod impact resistance of PPS changes little with temperature. The notched impact resistance values for Ryton R-4 are 1.5 ft · lb/in at −40°C (−40°F) and 1.0 ft · lb/in at 204°C (400°F).

As shown in Fig. 17.9, the compressive yield strength of Ryton R-4 decreases at temperatures above 93°C (200°F).

Tests on the Taber abrasion apparatus with a CS 17 abrasion wheel showed a weight loss of 0.034 gram (g) per 1000 revolutions for Ryton R-4.

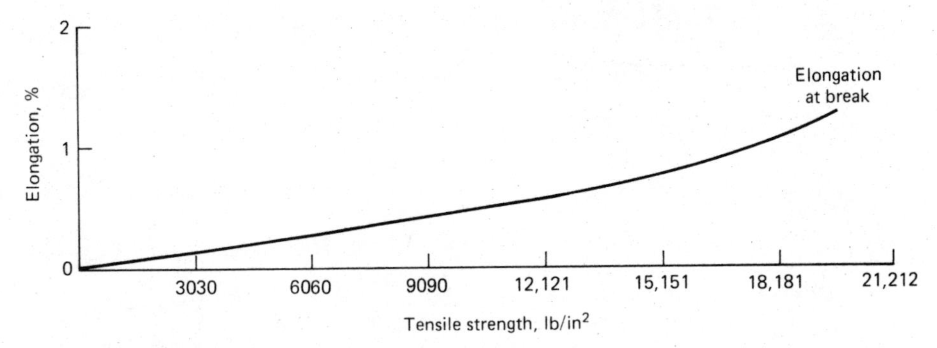

Figure 17.5 Stress-strain curve for Ryton R-4 at 73°F.

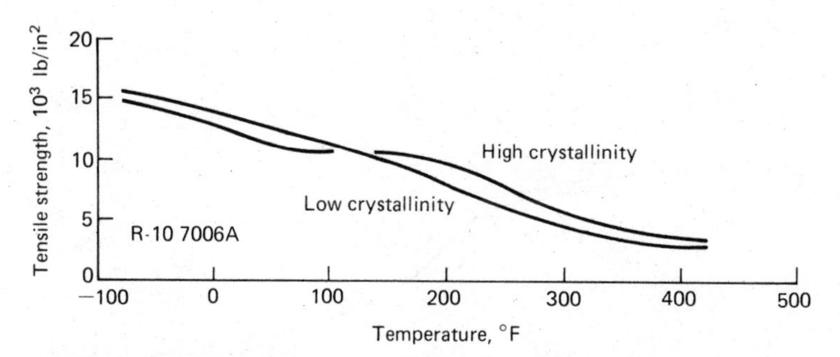

Figure 17.6 Effect of temperature on tensile strength of PPS.

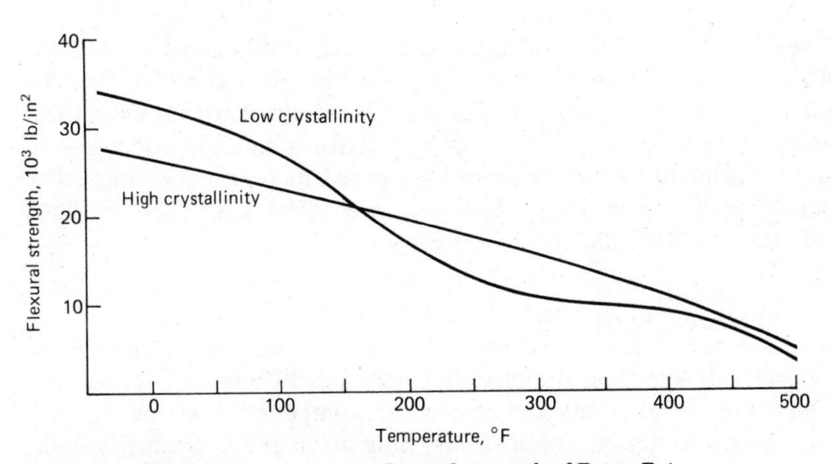

Figure 17.7 Effect of temperature on flexural strength of Ryton R-4.

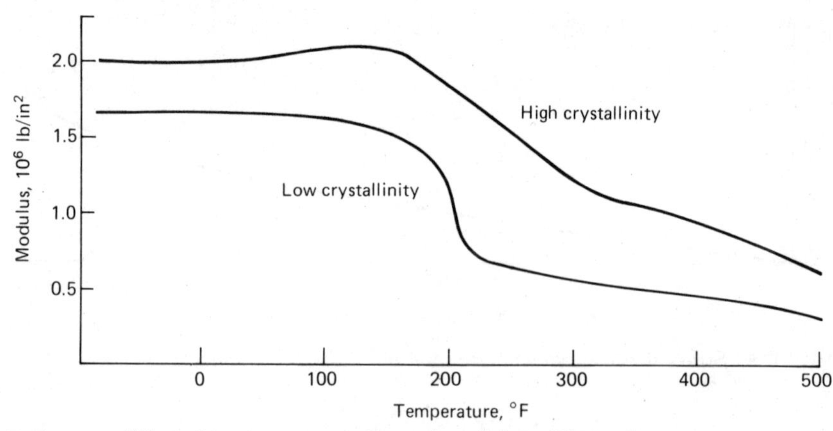

Figure 17.8 Effect of temperature on flexural modulus of Ryton R-4.

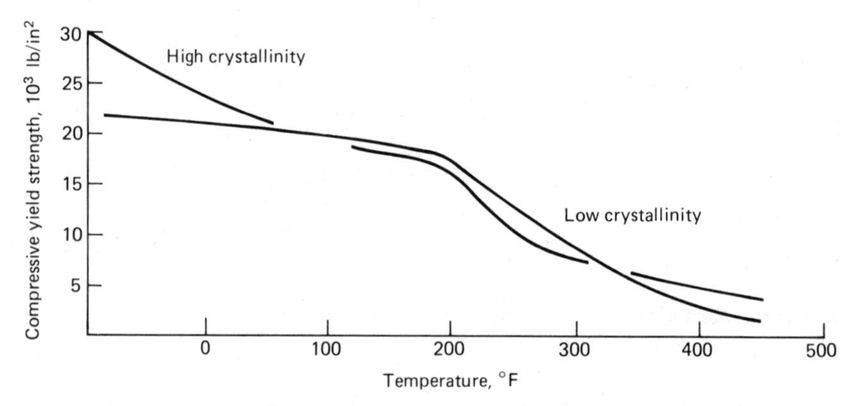

Figure 17.9 Effect of temperature on compressive yield strength of Ryton R-4.

The dielectric constant of Ryton R-4 is essentially constant at 23°C (73°F) up to 199°C (390°F) over a frequency range of 10 to 10^7 Hz. The dissipation factor at 23°C (73°F) also is essentially constant over a frequency range of 10 to 10^7 Hz. However, there is a definite decrease in the dissipation at higher temperatures over this frequency range. The volume resistivity of Ryton R-4 decreases from 2.5×10^{16} at 23°C (73°F) to 1.2×10^{12} at 199°C (390°F).

17.6 Applications of PPS

Because of its excellent dimensional stability, PPS is used for electrical and electronic sockets and connectors, component housings, switch relays, under-the-hood applications, hair drier grills, pumps, valves, capacitors, and high-performance film and filaments.

Its long-time use as reflectors for streetlight design and as essential components in the electronics industry is indicative of the usefulness of PPS. Less than 10,000 t of PPS was produced in the United States in 1987, but it is anticipated that this volume will increase to 20,000 t in 1992.

17.7 Glossary

amorphous Noncrystalline.

APC A PEEK prepreg.

Astrel A polysulfone.

carbonyl C=O

DTUL Heat deflection temperature under load.

ether group —O—

Fortron A PPS.

glass transition temperature The temperature at which brittle polymers become flexible as the temperature is increased.

Kadel A PEK.

hookean Complies with Hooke's law, i.e., stress is proportional to strain.

Hostatec A PEK.

Izod impact test A test of resistance to breaking when a weighted pendulum strikes a notched or unnotched cantilevered test bar.

modulus Stiffness.

PEEK Polyetherether ketone.

PEK Polyether ketone.

PES Polyether sulfone.

photooxidation Oxidation in the presence of sunlight.

polyetherether ketone PEEK.

polyether ketone PEK.

PPS Polyphenylene sulfide: $\left[-C_6H_4-S- \right]_n$

pultrusion The process in which a bundle of resin-impregnated filaments are pulled through a die.

Radel A polysulfone.

Ryton A PPS.

Stabar A PEEK film.

Stilan A PEK.

sulfone group —SO_2—

Taber abrasion test A test to determine wear resistance of a polymer.

T_g Glass transition temperature.

Udel A polysulfone.

UL Underwriters' Laboratory.

Ultrapek A PEK.

Victrex A PEEK, also PES.

Zyex A PEEK yarn.

17.8 References

Brody, D.: p. 86 in J. N. Epel, J. M. Margolis, S. Newman, and R. B. Seymour (eds.): *Engineering Materials Handbook,* vol. 2, ASM International, Metals Park, Ohio, 1988.

MacDermott, C. P.: *Selecting Thermoplastics for Engineering Applications,* Marcell Dekker, Inc., New York, 1984.

Margolis, J. M.: *Engineering Thermoplastics,* Marcell Dekker, Inc., New York, 1985.

May, R.: in "Polyetheretherketones," in J. I. Kroschwitz (eds.): *Encyclopedia of Polymer Science and Engineering,* vol. 12, Interscience Publishers, John Wiley & Sons, Inc., New York, 1988.

Rose, J. B., Clendinning, A. G. Farnham, R. N. Johnson, M. E. Sauers, L. A. McKenna, C. N. Merriam, and H. W. Hill: in "Polyetherketone," in R. B. Seymour and G. S. Kirshenbaum (eds.): *High Performance Polymers: Their Origin and Development,* Elsevier Science Publishing Co., Inc., New York, 1986.

Seymour, R. B.: *Polymers for Engineering Applications,* ASM International, Metals Park, Ohio, 1986.

Chapter

18

Liquid-Crystal Polymers

18.1 Introduction

Liquid crystals were observed by Reinitzer in 1888, but these unique products were not investigated, to any great extent, until the 1940s, when emphasis was on lyotropic (capable of forming an ordered solution) biopolymers, such as collagen. In 1956, Nobel laureate Paul Flory proposed that rigid rodlike polymers would retain their conformations in the liquid state. In 1965, S. Kwolek of Du Pont observed anisotropic solutions of aromatic PAs in alkylamide solutions.

Lyotropic aromatic PAs (Kevlar), melt-processible aromatic polyester (Ekkcel), and polyazomethines were developed for commercial applications in 1965, 1972, and 1974 by Du Pont, Carborundum, and Eastman Kodak, respectively. Some of J. Economy's technology was transferred from Carborundum to Dart in 1981, and Xydar then became the first American moldable liquid-crystal engineering polymer.

Schiff base, high-modulus, anisotropic melts of aromatic polyazomethine were produced by P. Morgan of Du Pont in 1977. Other investigators introduced tractable aromatic thermotropic polyesters by synthesizing polymers with bent, swivel, crankshaft, and ring-substituted structures as shown in Fig. 18.1.

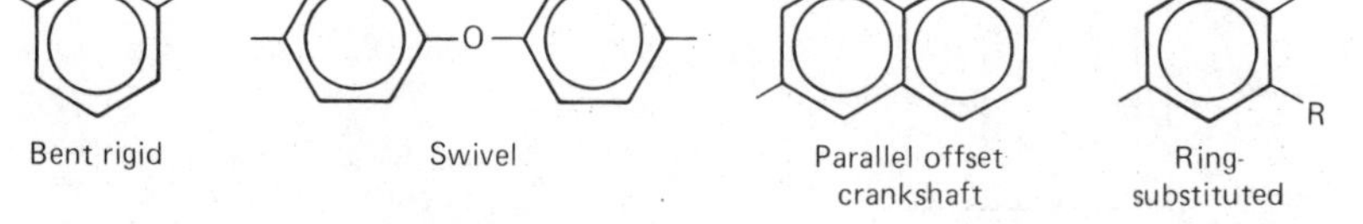

Figure 18.1 Structures of thermotropic polyesters.

18.2 Commercial Liquid-Crystal Polymers

The family of thermotropic polyesters derived from 2,6-dihydronaphthalene are called Vectra by Hoechst Celanese. These liquid-crystal polymers (LCPs) consist of relatively flat repeating units which are present as parallel, closely packed, fibrous chains in injection-molded parts. ICI supplies LCPs (self-reinforcing polymers) under the trade name Victrex SRP.

Because of ordered structures, LCPs have low melt viscosities and the readily molded products have exceptional good physically properties. LCPs have outstanding thermal oxidative stability and have a UL rating of 116°C (240°F). LCPs have an LOI of 42, are inherently flame-retardant and intumescent, and have a UL 94 rating of V-0. These engineering polymers are available as unfilled and fiberglass or carbon-fiber-filled composites. Over 9980 t of LCPs was produced in the United States in 1987.

The morphology of LCPs is illustrated in Fig. 18.2, which is based on ICI literature.

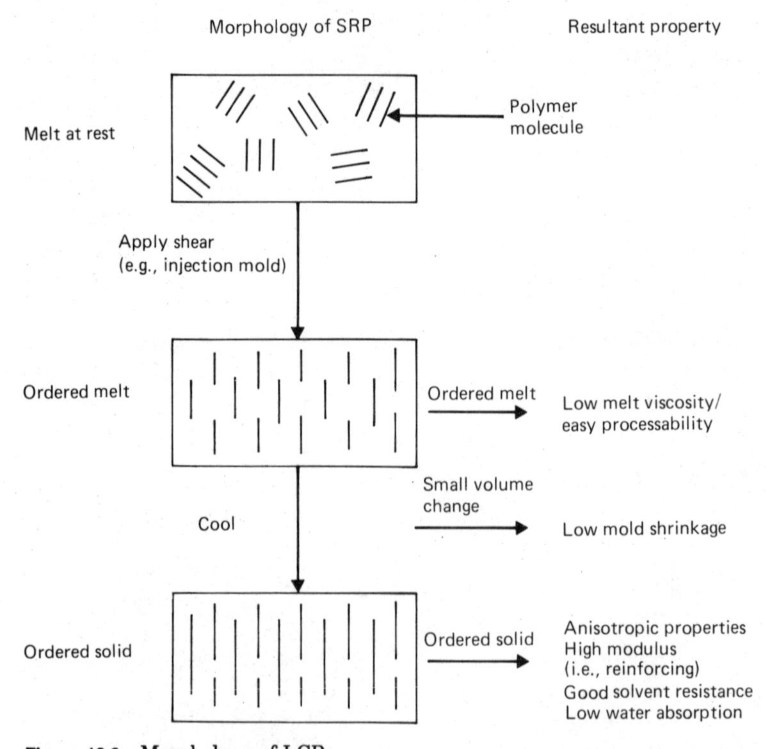

Figure 18.2 Morphology of LCPs.

18.3 Properties of LCPs

As was the case with fiber-filled extrusions, LCPs are anisotropic, i.e., their properties are enhanced along with the direction of flow and decreased across the flow direction. This anisotropic ratio is reduced by the presence of fillers. This is in contrast to conventional polymer composites, i.e., the properties of injection-molded LCPs, such as shrinkage, modulus, and creep are less anisotropic than those of the unfilled LCP.

Actually, because of the low melt viscosity of LCPs, thin sections can be molded with highly filled (70%) LCPs. The addition of bulk lubricants reduces the melt viscosity to a relatively low value. LCP moldings have good dimensional stability, low mold shrinkage, a low coefficient of thermal expansion, and little warpage, as is the case for many other engineering plastics. As shown in Fig. 18.3, the dynamic storage modulus of commercial LCPs, such as SRP 1500G, decreases dramatically at about 104°C (220°F). SRP 2300G behaves similarly, and the moduli of SRP 2300G and SRP 2300GL30 decrease gradually at lower temperatures.

As shown in Fig. 18.4, LCPs have good resistance to creep in the flow direction.

The effect of time on strained LCPs is plotted in Fig. 18.5.

Since these self-reinforcing polymers are crystalline, the DTUL of unfilled SRP 1500G is increased from 170 to 231°C (338 to 448°F) by the addition of fiberglass. These LCPs also retain their excellent me-

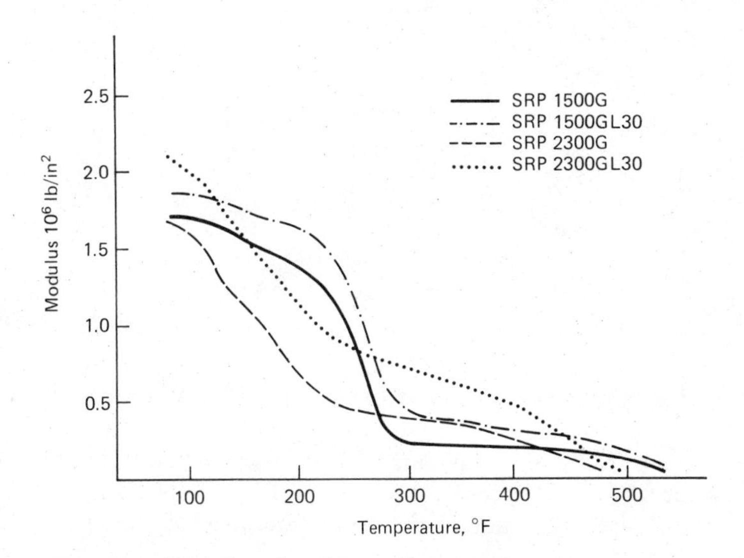

Figure 18.3 Variation of modulus with temperature.

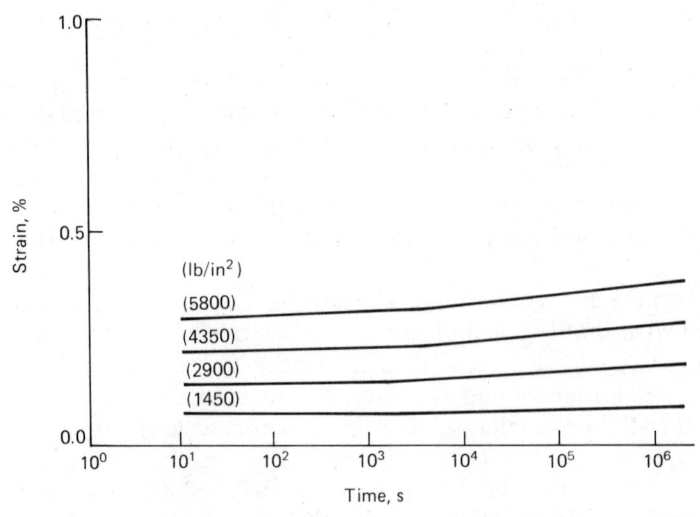

Figure 18.4 Interpolated creep data for Victrex SRP 1500GL30 at 73°F (23°C) (in flow direction).

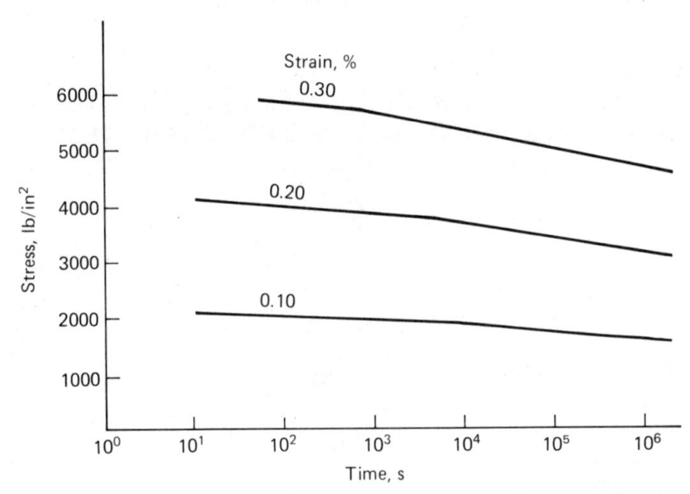

Figure 18.5 Isometric data for Victrex SRP 1500GL30 at 73°F (23°C) (in flow direction).

chanical properties when used for long periods of time at elevated temperatures.

These plastics have LOI values in the order of 36 percent and have a flammability rating of V-O according to the UL 94 test. The smoke emission of burning LCP is low and in the same order as that of PTFE, PC, and PF.

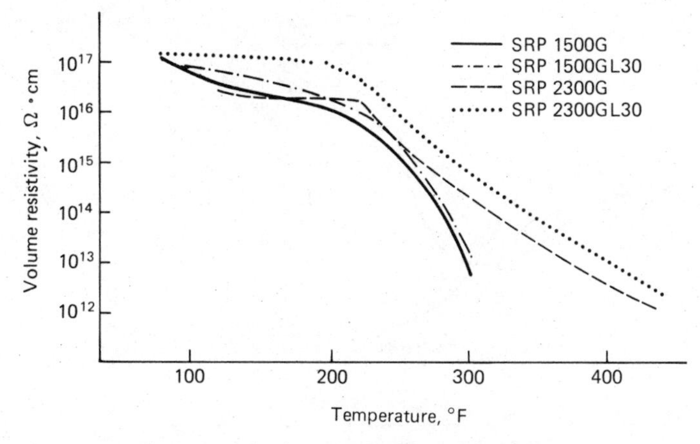

Figure 18.6 Variation of volume resistivity with temperature.

As shown in Fig. 18.6, the volume resistivity of SRPs decreases significantly at about 110°C (230°F).

As shown in Fig. 18.7, the dielectric constant (permittivity) of SRP 2300G and SRP 2300GL30 is essentially unaffected by an increase in temperature, but permittivity for SRP 1500G and SRP 1500GL30 increases abruptly at about 135°C (275°F).

As shown in Fig. 18.8, the effect of temperature on the loss tangent is similar to that of permittivity.

The effect of increasing the frequency on the loss tangent is shown in Fig. 18.9.

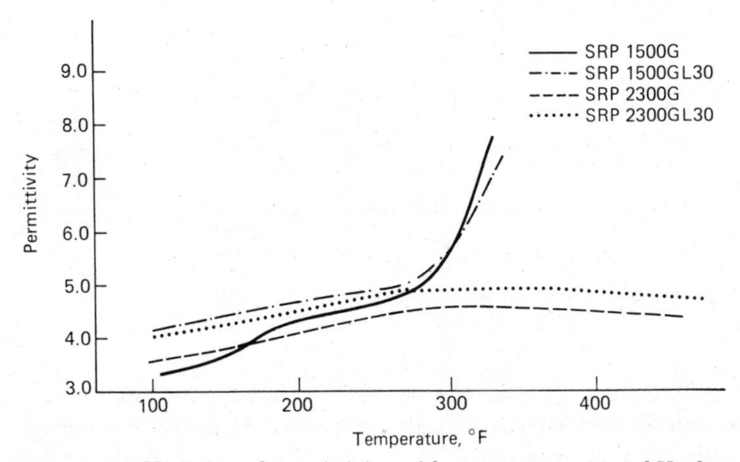

Figure 18.7 Variation of permittivity with temperature (at 1 kHz frequency).

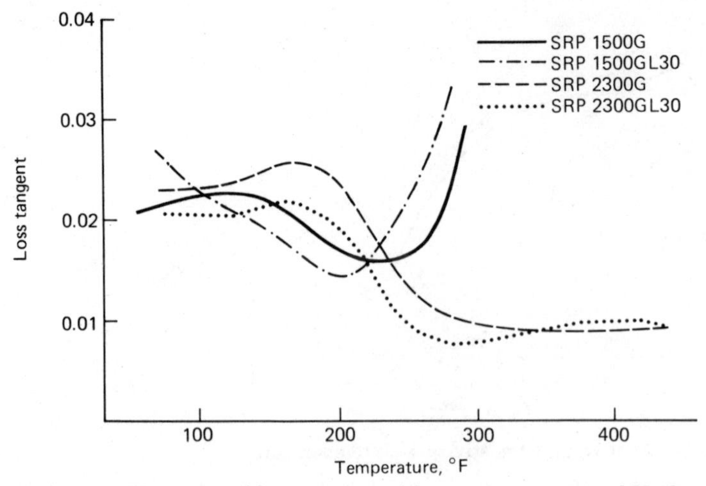

Figure 18.8 Variation of loss tangent with temperature (at 1 kHz frequency).

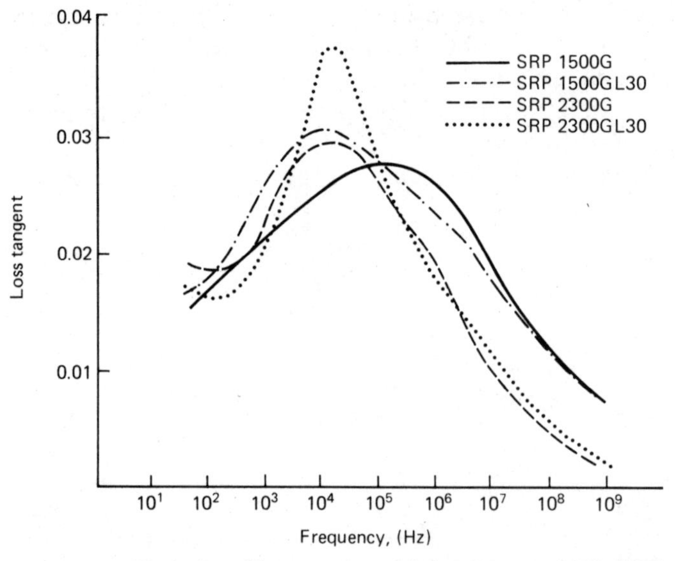

Figure 18.9 Variation of loss tangent with frequency at 23°C (73°F).

LCPs have excellent weatherability characteristics and are resistant to the deleterious effects of gamma radiation up to a level of 1,000,000 rad.

The properties of commercial LCPs are summarized in Table 18.1.

TABLE 18.1 Properties of Typical Liquid-Crystal Polymers

Property, units*	LCP	LCP + 40% fiberglass
Melting point T_m, °F	770	770
Glass transition temp. T_g, °F	—	—
Processing temp., °F	725	725
Molding pressure, 10^3 lb/in^2	10	10
Mold shrinkage, 10^{-3} in/in	4	2
Heat deflection temp. under flexural load of 264 lb/in^2, °F	650	600
Maximum resistance to continuous heat, °F	600	575
Coefficient of linear expansion, 10^{-6} in/(in · °F)	15	7
Compressive strength, 10^3 lb/in^2	6	10
Izod impact strength, ft · lb/in of notch	3.0	1.5
Tensile strength, 10^3 lb/in^2	18	13.5
Flexural strength, 10^3 lb/in^2	19.5	20.5
Percent elongation	3	1.8
Tensile modulus, 10^3 lb/in^2	2100	1870
Flexural modulus, 10^3 lb/in^2	1850	1300
Rockwell hardness	R61	R79
Specific gravity	1.35	1.70
Percent water absorption	0.1	0.1
Dielectric constant	3	3
Dielectric strength, V/mil	620	510
Resistance to chemicals at 75°F:†		
Nonoxidizing acids (20% H_2SO_4)	S	S
Oxidizing acids (10% HNO_3)	S	S
Aqueous salt solutions (NaCl)	S	S
Polar solvents (C_2H_5OH)	S	S
Nonpolar solvents (C_6H_6)	S	S
Water	S	S
Aqueous alkaline solutions (NaOH)	S	S

*lb/(in^2 · 0.145) = KPa (kilopascals); ft · lb/(in of notch · 0.0187) = cm · N/cm of notch.
†S = satisfactory; Q = questionable; U = unsatisfactory.

18.4 Theoretical Considerations of LCPs

The tendency to form liquid crystals is dependent on intermolecular repulsions which limit the number of rodlike chains that can occupy the available volume in a solution or melt. However, the melt is preferred for liquid-crystal formation, and rodlike molecules are preferred over random-coiled polymers in LCPs.

Aromatic PAs (aramids) have rodlike structures and form LCPs because of the presence of extended rigid chains. These aramids dissolve in dimethylacetamide in the presence of 5% lithium chloride (LiCl), in concentrated sulfuric acid, and in hydrogen fluoride (HF) to form lyotropic solutions. However, because of the insolubility of the stiff chain ester groups, very few polyesters form lyotropic solutions.

The formation of LCPs on melting (thermotropism) is related to the presence of stiffening groups (mesogens). Thermotropic LCPs exhibit birefringence or opalescence at temperatures above T_m.

Thermotropic aromatic polyester melts, which are usually nematic (in a parallel array), are formed when the chain stiffness can be maintained while the T_m is depressed. Thermotropic melts of aromatic polyesters are readily extruded or melt spun at temperatures just below T_m.

Poly-1,4-benzoate (Ekonol) can be sintered but cannot be molded by conventional techniques. However, Ekkcel or Xydar, which is a copolymer of 4,4′-biphenol and terephthalic acid, can be readily injection-molded. This statement also applies to Vectra, which consists of poly-1,4-benzoate modified by 2-oxynaphthalene-6-carbonyl units. Both Xydar and Vectra can be recycled and remolded.

Polyazomethines are produced by the condensation of terephthalaldehyde and methyl-4,4′-phenylenediamine. These LCPs are present in the nematic phase in lyophilic sulfuric acid solutions.

18.5 Applications

Because of their ease of molding, flame resistance, and thermal stability, LCPs are used as replacements for complex shapes, metals, and ceramic parts in the electronic, aerospace, chemical process, and appliance industries. About 350 t of LCP is used annually in the United States for electronics and telecommunications, but the major outlet at the present time is for molding the Ultra 21 cookware line of Tupperware. Over 1500 t of LCP is used annually in the United States for this cookware.

18.6 Glossary

amide group —$CONH_2$

anisotropic A system with ordered domains.

aramid Aromatic nylon.

aromatic Belonging to the benzene class, i.e., cyclic rather than straight-chain molecules.

cholesteric A nematic system with a superimposed spiral arrangement of nematic layers.

conformation Shape.

creep A change in dimensions over a long period of time.

dimethylacetamide $H_3CCON(CH_3)_2$

dynamic storage modulus The modulus measured under simulated load.

Ekkcel A polyester.

Ekonol A poly(1,4-benzoate).

ester group —COOR

Flory, P. A Nobel laureate and polymer scientist.

Hz Hertz.

intumescent A polymer which swells when heated or combusted.

isotropic A disordered system.

Kevlar An aramid.

LCP Liquid-crystal polymer.

liquid-crystal (LC) An ordered state which is stable at a temperature range between the solid state and an ordered fluid system.

LOI Limiting oxygen index, i.e., the maximum percentage of oxygen in a nitrogen-oxygen mixture in which the test sample does not support combustion via a candlelike flame.

loss tangent The ratio of the power loss to the total power transmitted, or the loss angle, i.e., imperfections of a dielectric.

lyotropic Capable of forming an ordered solution.

mesogenic Favoring the formation of liquid crystals.

mesomorphic Capable of forming crystals.

modulus Stiffness.

nematic Having a parallel array of polymer chains.

permittivity Dielectric constant.

polyazomethine A polymer produced by the condensation of terephthalaldehyde and methyl-1,4-phenylenediamine.

s Seconds.

smectic A nematic system assisted by end-group alignment.

SRP Self-reinforcing polymer, i.e., LCPs.

thermotropic Capable of forming an ordered melt.

T_m Melting point.

Tupperware Kitchenware sold by Dart.

UL Underwriters' Laboratory.

Vectra An LCP.

Victrex SRP LCPs.

Xydar A polyester LCP.

18.7 References

Blumstein, A. (ed.): *Liquid Crystalline Order Polymers,* Academic Press, Inc., New York, 1978.
Braun, G. H. (ed.): *Advances in Liquid Crystals,* Academic Press, Inc., New York, 1979.
Calundann, G. W., and A. C. Griffin: in "Thermotropic Polyesters," in R. B. Seymour and G. S. Kirshenbaum (eds.): *High Performance Polymers: Their Origin and Development,* Elsevier Science Publishing Co., Inc., New York, 1986.
Ciferri, A., W. L. Kiegbaun, and R. B. Meyer: *Polymer Liquid Crystals,* Academic Press, Inc., New York, 1982.
Gordon, M.: *Liquid Crystal Polymers,* Springer-Verlag New York, Inc., New York, 1984.
Kwolek, S. L., P. W. Morgan, and J. R. Schaefgen: in "Liquid Crystal Polymers," in J. I. Kroschwitz (ed.): *Encyclopedia of Polymer Science and Engineering,* vol. 9, Interscience Publishers, John Wiley & Sons, Inc., New York, 1987.
McChesney: p. 179 in J. N. Epel, J. M. Margolis, S. Newman, R. B. Seymour (eds.): *Engineering Materials Handbook,* vol. 2, AMS International, Metals Park, Ohio, 1988.

Chapter

19

Advanced Polymer Composites

19.1 Introduction

Biopolymeric composites, such as feathers and wood, have been available for thousands of years and fiberglass-reinforced polyester resins (GRP, FRP) have been in use since the early 1940s. The properties of these and other reinforced thermosets were discussed in Chaps. 2, 3, and 4, and other composites of specific engineering polymers have been discussed in subsequent chapters. The emphasis in this chapter is on advanced thermoplastic composites which have not been discussed previously in this book.

The sales of advanced thermoplastic composites in the United States exceeded $700 million in 1988 and should exceed $1 billion ($10^9$) by 1995. The present emphasis is on carbon-fiber-reinforced thermoplastics as replacements for metals in aircraft, but these successful applications will catalyze a large number of less strategic uses.

19.2 Fibrous Reinforcements

Fiberglass continues to be the reinforcement of choice for most composites. Since good interfacial adhesion is essential, new coupling or sizing agents and filaments with smaller diameters have been introduced. The interface of resins and carbon fibers, which have a higher modulus and lower density than glass fibers, has been enhanced by short-time (50 s) air oxidation at 760°C (1400°F), by short-term (25 s) anodic oxidation in aqueous sodium hypochlorite (NaOCl) at a pH of 10, by immersion in nitric acid (HNO_3), potassium permanganate ($KMnO_4$), sulfuric acid (H_2SO_4), and chromic acid (H_2CrO_7), and by treatment with undisclosed proprietary systems. Impact strength has been improved by the use of fiber blends (hybrid fibers) and by the use of longer fibers with smaller diameters, i.e., with higher aspect ratios.

19.3 Pultrusion

Composites with short fibers may be extruded or injection-molded, but the advantages of long-fiber reinforcements are lost unless filament winding or pultrusion techniques are used. In the pultrusion process, which is used for both thermoset and thermoplastic resins, an anisotropic composite is produced when the filaments impregnated with resin are pulled through an orifice. The pultrusion technique yields an unusually high ratio of reinforcement/resin in these unidirectional composites (UDC). The pultrusion process, which produces profiles of constant cross section and infinite length, may be modified by a pulforming process in which the composite is shaped into specifically designed shapes.

Thin-walled profiles and coated wire or cable can be produced by the pultrusion process. The composite can also be designed so that different filament/resin ratios are present in the same profile.

19.4 Design Considerations

The stress-strain curves of ceramics and metals are essentially linear (hookean) at moderate strain, but the linear portion of the stress-strain curves for polymers is usually limited to very low stresses. Fortunately, this portion of the stress-strain curve can be amplified in the fiber direction better than in the transverse direction.

The law of mixtures may be applied to the linear portion of the stress-strain curve. This law states that the modulus of the composite, E_c, is equal to the sum of the volume fraction of the reinforcing fiber, V_f, multiplied by its volume fraction E_f, plus the modulus of the matrix, E_m, multiplied by its volume, $1 - V_f$.

$$E_c = V_f E_f + (1 - V_f)E_m$$

The law of mixtures may also be used for approximations outside the linear portion of the stress-strain curve. A stress-strain diagram for a typical composite is shown in Fig. 19.1.

Since the stress-strain curves for anisotropic composites are not linear, the law of mixtures is not particularly useful for determining the transverse properties of these UDCs. UDCs under longitudinal tensile load may fail by a brittle fracture, fiber pullout, and debonding interfacial shear failure. This failure under longitudinal compressive load

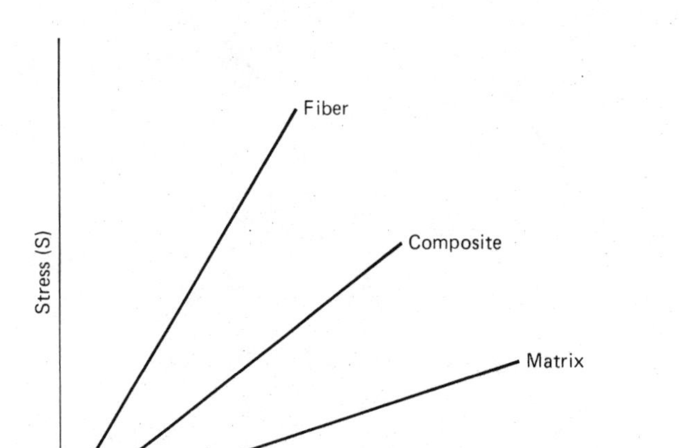

Figure 19.1 Stress-strain diagram for a typical thermoplastic composite.

may be via transverse tensile failure, microbuckling of fibers, and shear failure.

The principal modes of failure under transverse tensile load are failure of the matrix, interfacial debonding, and/or fiber splitting. The principal mode of failure under transverse compression load may be accompanied by shear failure of the matrix, interfacial debonding, and/or fiber crushing. The principal mode of failure under in-plane shear load is shear failure of the matrix, which may or may not be accompanied by interfacial debonding.

The properties of a thermoplastic composite are inversely proportional to the diameter of the fibers and directly proportional to the length and the concentration of the fibers in a composite. As shown by Fig. 19.2, the tensile strength of the composite increases linearly with the increased volume of fiber, providing the matrix has elastic behavior. The increase is lessened, to some extent, but the curve is linear if the matrix is not elastic. The maximum fiber content of a unidirectional composite is 90.7 percent for hexagonal packing and 78.5 percent for square packing. However, it is not practical to obtain fiber volumes greater than 70 percent in commercial composites.

As shown by Fig. 19.3, the tensile modulus of composites increases as the fiber length increases up to a fiber length of about 2 in, at which the fiber tends to act as a continuous fiber. However, the strength of the resin matrix is independent of the fiber length or fiber content.

The strength of the interfacial bond is decreased by the presence of voids, and this decrease is related linearly to the void volume.

In general, creep of composites is inversely proportional to the con-

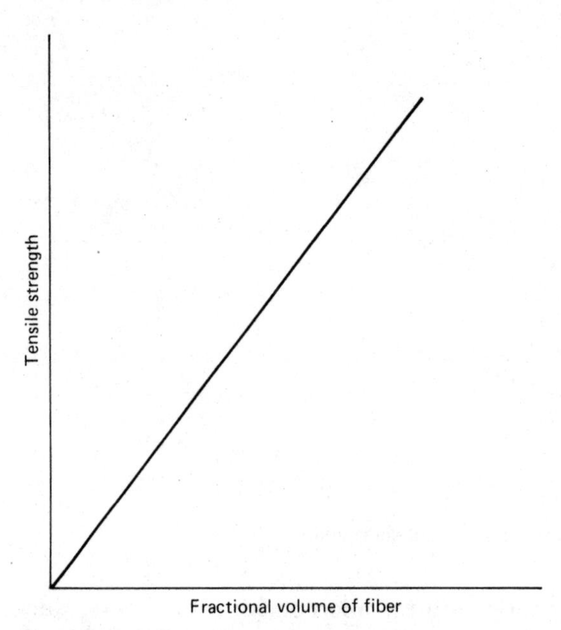

Figure 19.2 Effect of fractional volume of fiber on tensile strength.

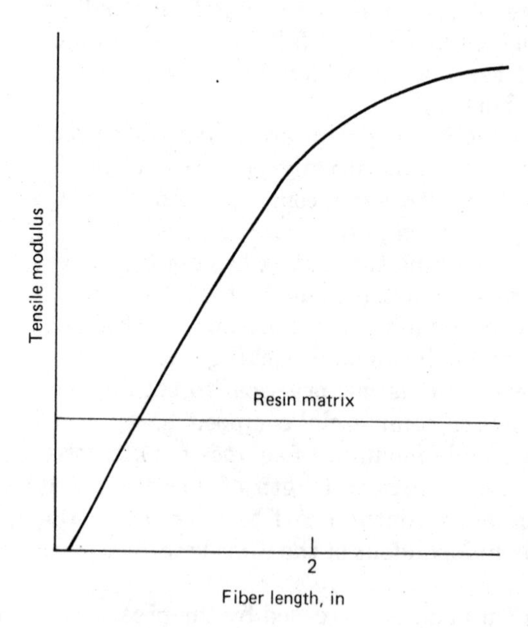

Figure 19.3 Effect of fiber length on tensile modulus.

centration of fibrous reinforcement. For example, it is not uncommon for the strain to decrease by 30 percent when 30 percent fiberglass is added to a thermoplastic polymer. A decrease in interfacial adhesion, such as that which occurs in aqueous environment, reduces the resistance to creep and fatigue. This decrease is also enhanced by an increase in temperature and exposure to UV radiation.

High aspect ratios of the reinforcing fiber or filler enhance the resistance of a composite to fatigue. However, there is an upper limit of the aspect ratios in the 500 range.

19.5 Rheology of Composites

Liquids are usually newtonian, i.e., the stress S, as measured by the deformation, is proportional to the rate of deformation (flow), $d\gamma/dt$, where the proportionality constant η is the viscosity coefficient. The symbol $\dot{\gamma}$ is sometimes used for the rate of deformation.

$$S = \eta\, d\gamma/dt = \dot{\gamma}$$

Deviations from newtonian behavior occur as the molecular weight increases and as the magnitude of the interfacial bond and the size of pendant groups and branches increase. However, good flow characteristics are noted for melts with good geometry, such as LCPs. Newtonian flow and some of the classic deviations from this flow are shown in Fig. 19.4.

The pressure used in injection molding or extrusion is inversely proportional to the temperature. There is also a phenomenon called the Weissenberg or Barus effect, which is related to die swelling and melt fracture of extrudates. These effects are more evident in polymeric composites.

19.6 Viscoelasticity

The viscous component is important in the processing of both filled and unfilled thermoplastics. However, the elastic component of viscoelastic plastics, as shown by Hooke's law, is the dominant component. Hooke's law states that

$$S = G\gamma$$

where S is stress, G is modulus, and γ is deformation.

However, both hookean and newtonian components must be considered in solid plastics. The stress-time plots may be represented by appropriate combinations of the Maxwell and Voigt-Kelvin models, which are shown in Figs. 19.5 and 19.6. These models include a

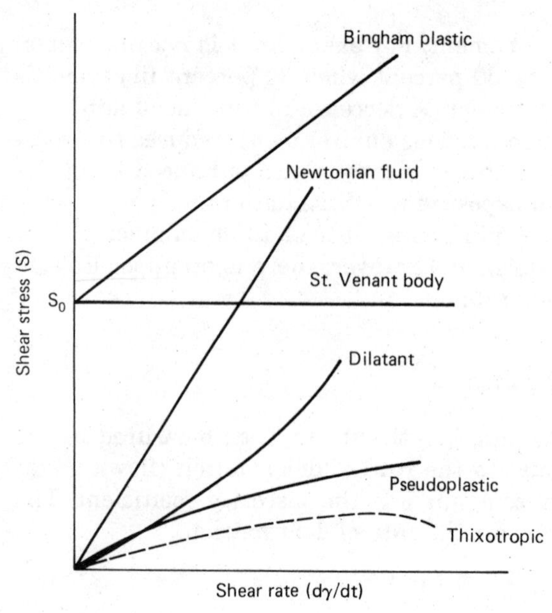

Figure 19.4 Flow behavior of polymers.

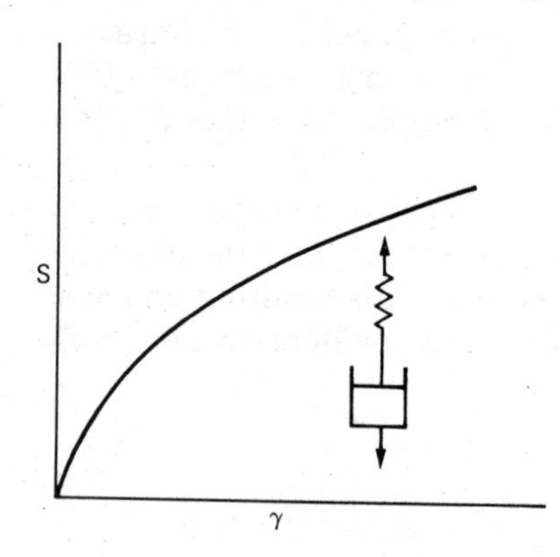

Figure 19.5 Stress-strain plot for stress relaxation in the Maxwell model.

weightless or ideal elastic spring with a modulus of G and a (fluid) dashpot or shock absorber with a viscosity of η.

Thus, while the flow of hot polymer melts is pseudoplastic, this flow approaches newtonian flow above a stress yield of S_o. This Bingham plasticity is shown in Fig. 19.4.

For simplicity, polymers are considered to be isotropic; in this type of polymer the principal force is shear stress S. Of course, composites are usually anisotropic and at high stresses also have stresses perpendicular to the plane of shear stress as evidenced by die swell.

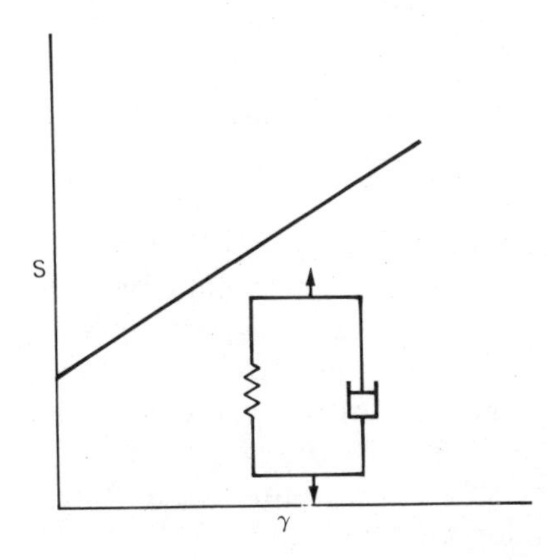

Figure 19.6 Stress-strain plot for a Voigt-Kelvin model.

As shown in Fig. 19.7, there are several subclasses of viscoelasticity which are temperature- and time-dependent, i.e., viscous glassy (below T_g), leathery, rubbery, rubberlike, and viscous flows. Any elongation in the viscous glass region is essentially hookean and reversible and dependent on the bending, stretching, and deformation of covalent bonds. The polymer becomes somewhat flexible or at least less brittle at temperatures slightly above T_g and has reversible rubbery time-dependent flow in the leathery-rubbery or viscoelastic region.

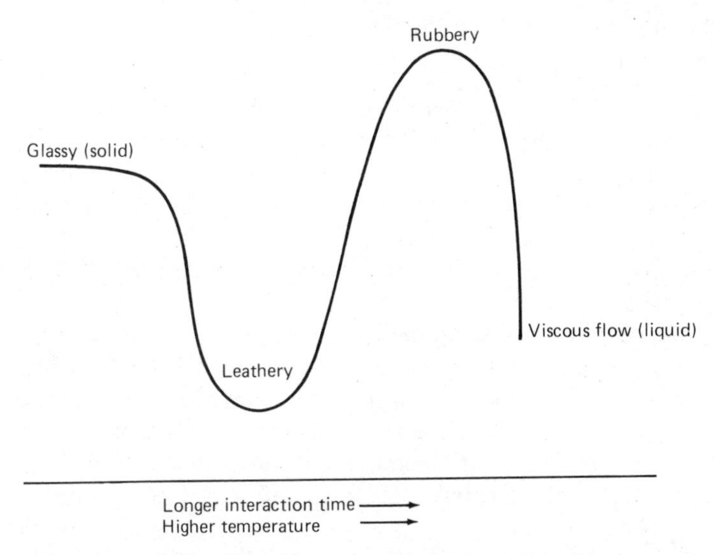

Figure 19.7 Regions of material response as a function of interaction (reaction) time and temperature.

19.7 Effects of Temperature

The flow is restricted somewhat by intermolecular forces and/or reinforcements in the rubberlike regions and exhibits irreversible flow in the viscous flow or liquid region.

The physical properties of general-purpose plastics decrease at temperatures above 93°C (200°F), but most advanced thermoplastic composites can be used for long time intervals at much higher temperatures. The tensile strength and tensile modulus of both unfilled and reinforced high-performance thermoplastics decrease linearly as the temperature is increased, but the slope is less for the polymeric composites than for the unfilled polymers in the same temperature range.

The heat deflection temperature under a load of 264 lb/in^2 may be used as a guide for the resistance of composites to elevated temperatures. UL provides temperature index values which are based on empirical heat gauging procedures, and these can be used as ceiling temperatures for continuous service.

19.8 Polymers for Engineering Applications

The trade names and producers of engineering polymers are shown in Table 19.1.

19.9 Unique Applications of Polymeric Composites

A prestressed fiberglass polyester bridge was constructed in Düsseldorf from 59 tendons based on 19 rods of 60,000 polyester-impregnated glass fibers. The world's largest pultrusion profile measuring 62 ft × 8 ft × 2.5 in was used to clad buildings in Canada's defense warning system. Sergei Bubka set a new world's record for vaulting using an FRP pole. FRP is also being used instead of amalgams for tooth filling.

Doors that do not warp are being molded from FRP. The use of composites in appliances is expected to double to 85,000 t in 1990. BMC polyester is being used as food containers for ovenable dishes. The winning boat in the National Kayak and Canoe Marathon Racing Championship held at Bowling Green, Kentucky, in 1987 was an extruded-chain polyethylene-reinforced (Spectra) vinyl ester kayak, which weighed 24 lb.

The success of the Pontiac Fiero FRP automobile body has catalyzed many other uses of FRP in automobiles. The tilt hood assembly of

TABLE 19.1 Principal Producers of Engineering Plastics

Company	Trade name	Composition
Allied Signal	Capron	Nylon 6
Amoco Chemical	Torlon	PAI
Asahi	Xyron	Modified PPE
BASF	Ultramid	Nylon 6
Bayer AG	Durethan	Nylon 6-PU
Borg-Warner Chemical	Prevex	PPE-PS blend
Dynamit Nobel AG	Trogamid T	Nylon 6
Du Pont	Kapton	PI
	Rynite	Polyester
	Selar	Nylon-polyolefin blend
	Zytel	Nylon 6,6
	Delrin	Acetal homopolymer
	Minlon	Nylon 6,6
Eastman Kodak	Kodar	Polyester
Emser	Amidel	Nylon 11
GE	Ultem	PEI
	Valox	Polyester
	Lexan	PC
	Noryl	PPO-PS blend
	—	PPO
	—	Polyester carbonate
	Xenoy	PBT-PC blend
ICI	Victrex	PEEK
Mitsubishi Gas and Chemical Co., Ltd.	Iupilon	PC
3M	Iupace	Modified PPE
	Astrel	Polysulfone
Mobay	Merlon	PC
	Bayblend	ABS-PC blend
Monsanto Polymer Products	Vydyne	PA 6
Phillips Petroleum	Ryton	PPS
Polyplastics	Duracon	Acetal copolymer
Rohm & Haas	Kamax	PI
Teijin	Panlite	PBT
Ticona Polymerwerke AG	Hostaform	Acetal
Ube	Upilex	PI
Unitika Chemicals	Unitika	Polyarylate

Ford trucks is one of the largest SMC moldings to date. Obviously, SMC will be the choice for automobiles and vans in the future. SMC is now being used as panels on Greyhound buses.

Glass-fiber-reinforced foamed PPS is being used for automotive manifolds. According to a report from Springborn Laboratories, the amount of plastics in automobiles will increase from 220 lb per car in 1987 to 275 lb in 1997. Pontiac abandoned the production of the Fiero because of problems not associated with plastics.

The success of the Voyager mission was a dramatic affirmation of the ability of plastics to outperform conventional materials. Burt Rutan, designer and builder of the Voyager, received the Society of Plastics Engineers (SPE) 1987 Unique and Useful Industrial Plastics Product Award. Present-day communication satellites consist of more than 4000 plastic composite parts (APC) and have 170 times the capability of "Early Bird," the first communication satellite which was launched in 1965. The weight savings by the use of APC instead of aluminum is 30 percent, and $6000/lb is saved by weight reduction. This increases to $27,000 in the upper extremities of space.

Likewise, the flight range of the Voyager was increased by 8 mi/lb reduction in weight. Filament-wound FRP has been tested as a principal component of the U.S. Air Force prototype X-29 airplane. A graphite-reinforced epoxy resin composite is being used in the experimental flaperons, flaperon seals, and fixed trailing edge assemblies of the V-22 Osprey tilt-rotor aircraft for the U.S. Marine Corps. The Monarch, a 44-lb FRP pedal-powered helicopter, is being tested at Santa Ana, California.

A General Electric jet engine duct, which was molded from a BMI resin composite, has passed the flame test of the Federal Aviation Administration (FAA). Northrup is using a graphite-reinforced PEEK composite for its F-5F landing gear strut door and access panel. Based on these "breakthroughs," it has been predicted that polymer composites will account for 90 percent of the structure of all aircraft, including engines, in the future.

As shown by the data in Table 19.2, most engineering polymers have acceptable flexural moduli at 204°C (400°F), and PEK and PEEK have modulus values of about 400 × 10^3 lb/in^2 at 260°C (500°F). The flexural moduli at these temperatures are much higher for polymeric composites.

TABLE 19.2 Flexural Modulus at Elevated Temperature for High-Temperature, Glass-Fiber-Reinforced Composites

Base resin	Glass content, %	Flexural modulus (lb/in^2 × 10^3)						
		73°F	200°F	300°F	350°F	400°F	450°F	500°F
PES	30	1260	1150	1120	1100	400	—	—
PEI	30	1380	1200	1150	1120	420	—	—
PPS	40	1900	1500	100	750	450	250	—
PEEK	30	1500	1450	950	550	350	350	350
PEK	30	1530	1480	1000	650	500	475	450

19.10 Future of Engineering Polymers

In spite of their high cost per pound, which is less significant when relative specific gravities are considered, the use of engineering plastics will continue to grow at an annual worldwide rate approaching 10 percent. The high cost of resin production will limit the number of producers, but because of a more realistic profit margin, which is not characteristic of general-purpose plastics, the well-established producers will be able to enhance their technical service capabilities.

The need for education, which has plagued the general-purpose-plastics industry, is even greater for the more complex engineering-plastics and composites industries. The future development of computers, robotics, and other sophisticated industries is dependent, to a large extent, on a transfer of knowledge of engineering polymers to the design engineers and inventors in these industries. The latent discovery of the piezoelectric properties of PVDC and the conductivity of polyacetylene has already resulted in quantum jumps for progress in the robotics and computer industry, and comparable applications of other characteristic properties of engineering polymers will have a comparable effect on other modern industries.

New innovations in engineering polymers and blends of these polymers are ensured. However, in addition to the transfer of knowledge to those engineers who can utilize this knowledge in new applications, there is also a need for additional advances in processing and applications. The development of extended chain polyolefins (Spectra) and PET soft drink bottles that are essentially isotropic should catalyze new developments that would be impossible in the absence of engineering polymer science and technology.

19.11 Glossary

η (eta) Coefficient of viscosity.

$\dot{\gamma}$ $d\gamma/dt$, the rate of deformation.

a A constant related to holes or free volume in polymers.

Andrade equation A modification of the WLF equation:

$$\log \frac{\eta}{\eta_g} = \frac{a(T - T_g)}{b + T - T_g}$$

anisotropic Properties dependent on the direction of the test.

Arrhenius equation $\eta = Ae^{E/RT}$

aspect ratio l/D = length/diameter.

b A constant related to holes or free volume in polymers.

Barus effect Die swelling.

Bingham equation $S - S_o = \eta \, d\gamma/dt$

Bingham plastic A plastic that does not flow until its threshold stress (S_o) is exceeded.

biopolymer A polymer from animals or plants, such as proteins, nucleic acids, cellulose, and starch.

carbon fiber A fiber produced by the pyrolysis of acrylic fibers or pitch.

composite A combination of two or more components, usually a resin and a reinforcing fiber.

coupling agent A compound containing one group which is attracted to a filler surface and another group which is attracted to the resin.

creep A change in the shape or size over a long period of time.

dashpot A piston in a cylinder containing a liquid with a viscosity of η.

die swell The expansion of an extrudate as it emerges from the extruded die.

dilatant Shear thickening.

DTUL Heat deflection temperature under a load of 264 lb/in^2.

elasticity The rapid reversible recovery of a stretched polymer.

fiber, hybrid A mixture of fibers.

fiberglass Extruded glass filaments.

Fiero An automobile with an FRP (GRP) body.

filament winding A process in which resin-impregnated filaments are wound on a rotating mandrel.

forces, intermolecular Attraction between molecules.

G Shear modulus.

graphite Carbon.

GRP Glass-reinforced plastic.

hookean Conforming to Hooke's law.

Hooke's law $S = G\gamma$

ideal spring A weightless spring with a modulus of G.

impact strength The resistance to brittle fracture.

interfacial adhesion Adhesion between the fiber (discontinuous phase) and its resin matrix (continuous phase) in a composite.

isotropic Having similar properties in all directions.

Mc The critical molecular weight at which chain entanglement begins.

matrix The resin.

Maxwell model An ideal spring and dashpot connected in series.

modulus Stiffness.

newtonian Conforming to Newton's law.

Newton's law $S = \eta d\gamma/dt$

pendant groups Functional groups attached to the principal polymer chain.

pH A scale of acidity; the degree of acidity decreases as one goes from 1 to 7, and the degree of alkalinity (basicity) decreases as one goes from 14 to 7.

piezoelectric Generating electric current from mechanical pressure.

power law $$S = K\left(\frac{d\gamma}{dt}\right)^n$$

pseudoplastic Shear-thinning.

pultrusion A process in which a bundle of polymer-impregnated filaments is pulled through an orifice.

PVDC Polyvinylidine chloride.

rheology The science of flow.

S Shear stress.

SMC Sheet-molding compound.

Spectra An extended-chain polyolefin.

strain Elongation.

stress The resistance to change in shape and size as measured by tensile-stress strength, etc.

Supic A PPS.

T_g Glass transition temperature.

thermoplastic A polymer which may be reversibly softened by heat.

thermoset A cross-linked polymer.

thixotropic Losing viscosity under stress.

UDC Unidirectional composite.

UV Ultraviolet.

viscoelasticity A combination of elastic and viscous behavior.

viscosity Resistance to flow.

Voigt-Kelvin model An ideal spring and dashpot connected in parallel.

Weissenberg effect Die swelling.

WLF Williams-Landel-Ferry equation.

19.12 References

Agarwal, B. D., and L. J. Broutman: *Analysis and Performance of Fiber Composites,* John Wiley & Sons, Inc., New York, 1980.
Broutman, L. J., and R. H. Krock (eds.): *Composite Materials,* Academic Press, Inc., New York, 1974.
Clagett, D. C.: in "Engineering Plastics," in J. I. Kroschwitz (ed.): *Encyclopedia of Polymer Science and Engineering,* vol. 6, Interscience Publishers, John Wiley & Sons, Inc., New York, 1986.
Delmonte, J.: *Technology of Carbon and Graphite Fiber Composites,* Academic Press, Inc., New York, 1974.
Dostal, C. A. (ed.): *Engineered Materials Handbook,* vol. 1, *Composites,* ASM International, Metals Park, Ohio, 1987.
Donnet, J. B., and R. C. Bansal: *Carbon Fibers,* Marcel Dekker, Inc., New York, 1984.
Eirich, F. R. (ed.): *Rheology,* Academic Press, Inc., New York, 1956.
Folkes, J.: *Short Fiber Reinforced Thermoplastics,* John Wiley & Sons, Inc., New York, 1982.
Hancox, N. L.: *Fiber Composite Hybrid Materials,* Applied Science Publishers, London, 1981.
Hull, D.: *An Introduction to Composite Materials,* Cambridge University Press, Cambridge, 1981.
Johnson, A. F.: *Engineering Design Properties of GRP's,* British Plastics Federation, London, 1979.
Liedtke, M. W., and W. H. Todd: *Advanced Composites,* ASM International, Metals Park, Ohio, 1985.
Lubin, I.: *Handbook of Composites,* Van Nostrand Reinhold Co., Inc., New York, 1982.
Moore, G. R., and D. E. Kline: *Properties and Processing of Polymers for Engineers,* Prentice-Hall, Inc., Englewood Cliffs, N.J., 1984.
Plueddemann, E. P. (ed.): *Interfaces in Polymer Matrix Composites,* Academic Press, Inc., New York, 1974.
Richardson, T.: *Composites: A Design Guide,* Industrial Press, Inc., New York, 1987.
Schwartz, M. M. (ed.): *Composite Materials Handbook,* McGraw-Hill Book Company, New York, 1984.
Schwartz, M. M. (ed.): *Fabrication of Composite Materials,* ASM International, Metals Park, Ohio, 1985.
Serafini, T. O.: *High Temperature Polymer Matrix Composites,* Noyes Publications, Noyes Data Corporation, Park Ridge, N.J., 1987.
Titow, W. V., and B. J. Landham: *Reinforced Thermoplastics,* Applied Science Publishers, London, 1985.
Tsai, S. W. (ed.): *Composite Materials,* ASTM Special Technical Publication 674, Philadelphia, 1979.
Ward, I. M.: *Mechanical Properties of Solid Polymers,* John Wiley & Sons, Inc., New York, 1971.
Watts, A.: *Commercial Opportunities for Advanced Composites,* American Society for Testing and Materials, Philadelphia, 1980.

Chapter

20

Design Considerations

20.1 Introduction

Since the first commercially available plastics, such as Celluloid, Bakelite, Glyptal, Plexiglas, PVC, PS, and PE, were seldom used as engineering plastics, there was little incentive for the producers to develop engineering design data on these materials. In contrast, considerable design data became available on metals as the science of metallurgy developed.

Nevertheless, since many of the newer plastics are used for engineering applications, the producers of these materials have developed considerable physical data which can be used by engineers for innovative design for aerospace, marine, automotive, electronic, and structural parts. Some design data have been provided for specific polymers in preceding chapters. Of course, many of these data can be used for other polymers than those discussed in those specific chapters. Additional information on design is provided in this chapter, which is the terminal chapter in this sourcebook.

20.2 Viscoelasticity

All polymers have the properties of viscous liquids, as well as the elasticity of solids. Thus, unlike elastic solids, polymers tend to creep irreversibly when placed under stress for long periods of time. This tendency to creep increases with temperature and when platicizers are present and decreases for reinforced polymers, hydrogen-bonded polymers, crystalline polymers, and thermosets.

Elastic-strain energy and crack-formation energy contribute to the total energy of viscoelastic polymers. Thus, a crack will spread throughout a polymer if the spreading lowers the total energy.

Elastic-strain energy decreases until the crack has spread throughout the entire polymer, i.e., the strain energy approaches zero. Fracture of viscoelastic materials is related to crack fracture and/or undetectable defects in the material.

20.3 Reinforced Polymers

Since phenolic, amino, and unsaturated polyester resins are extremely brittle, it has been customary to reinforce these thermosets with fibrous fillers. Since the properties of these thermosets, as well as those of vulcanized rubber, were improved dramatically by the addition of fillers, it was originally assumed that networks are essential for effective reinforcement.

PS was reinforced for marine applications in the 1950s, and reinforcements have been added to all thermoplastics since that time. It has been observed that such reinforcements are more effective with crystalline rather than amorphous polymers, but there is usually sufficient improvement to justify the reinforcement of amorphous engineering polymers. Fiberglass continues to be the major reinforcing filler, and its interfacial adhesion is improved by the presence of small amounts of coupling agents which form chemical bonds between these additives and the resin and fiber.

Talc and mica, which are crystalline platelets, are also used as reinforcing fillers. In contrast to the usual fiber-reinforced plastics, which are anisotropic, the composites containing these fillers tend to be isotropic.

20.4 Predictable Characteristics

In spite of many unsolved problems in design, there are several obvious characteristics which aid the designer in selection of engineering polymers. If transparency is essential, one usually chooses an unfilled amorphous plastic, such as PMMA, PC, polysulfone (PES), or polyarylates. The polyarylates also have excellent inherent resistance to degradation by UV radiation.

When toughness is measured by notched Izod impact, the toughest engineering polymer is super-tough nylon 6,6 which has an impact resistance of about 40 ft · lb/in of notch. PC also has a high impact resistance value of about 15 ft · lb/in of notch, but this property is dependent on the thickness of the specimen.

PPE, PC, nylon 6,6, PET, and PBT resins reinforced by 30 percent fiberglass have impact values of about 2 ft · lb/in of notch. Regardless of these comparable values, there is little correlation between the Izod impact ratings and the actual toughness. For example, multiple blows by a hammer will break PC, but nylon and acetals (POM) are resistant to multiple hammering blows. The fiberglass-reinforced engineering polymers, such as PPE, PET, PBT, and PC, and unfilled nylon 6,6 have high creep modulus values, but unreinforced super-tough nylon 6,6 has poor resistance to creep.

Nylons, PC, PES, PPE, and POM have high elongations under tensile stress, but these values are reduced to less than 5 percent when these polymers are reinforced by 30 percent fiberglass. In contrast, fiberglass-reinforced PET, PBT, PC, PPE, and nylon have high flexural strengths. PES, POM, PC, PPE, and ABS have flexural strengths greater than 12,000 lb/in^2.

Reinforcements also increase the thermal resistance of engineering polymers. Thus, the heat deflection temperature of fiberglass-reinforced PI, PEEK, arylates, nylon, PET, PBT, PES, and PC are all above 140°C (300°F). The heat deflection temperatures under a load of 264 lb/in^2 of unfilled arylates, PEEK, PI, POM, PC, PPE, and ABS are all above 100°C (212°F). Likewise, the moduli of fiberglass-reinforced PET, PBT, PPE, PC, and nylon are all above 1000 lb/in^2. The moduli of unreinforced POM, PES, ABS, PEE, and PC are in the 350 to 400 lb/in^2 range.

Engineering polymers have characteristic low specific gravity values in the range of 1.05 to 1.20, but these values are increased by the addition of fiberglass or mineral fillers. In contrast, the specific gravity can be decreased by the addition of hollow glass spheres (syntactic foam) or by foaming with a propellant gas. It should be noted that on a weight basis, structural foams are stronger than their solid counterparts.

20.5 Electrical Properties

Since most polymers are nonconductors, they are widely used in electrical and electronic applications. Nonconductivity is undesirable in housings for business machines, etc., because of electromagnetic interference (EMI). This problem can be solved by the application of conductive coatings, the incorporation of metal platelets, or the addition of antistats, which conduct current.

The dielectric strength values for reinforced and nonreinforced polymers are usually greater than 400 volts per mil (V/mil). For example, PEI and PPE have dielectric strengths of 710 and 550 V/mil, respectively. UL values for arc resistance show that POM has a value of 240. PEEK, PET, PES, and nylons all have values greater than 100, but the values for fiberglass-reinforced PC and PPE are about 10.

20.6 Ease of Processing

Since thermosets are processed as prepolymers, which are cross-linked after forming, they are readily processed and fabricated. Crystalline polymers, such as nylons, POM, PPS, and LCPs, as well as amorphous polymers, such as ABS, PPE, PC, and PES, are readily processed. Reinforced polymers are slightly more difficult to process than the corresponding unfilled polymers. Since absorbed moisture may make processing of some polymers more difficult, it is usually advantageous to dry the molding powder before molding or extruding. Automation usually simplifies the processing step.

20.7 Dimensional Changes in Processing

Since most plastics shrink when molded and expand when extruded, it is essential to make appropriate changes in dimensions when designing molded or extruded articles. In general, crystalline polymers, such as POM and nylons, shrink more than amorphous polymers in the molding process. This shrinkage can be reduced dramatically by the addition of fillers or reinforcing fibers. However, reinforced moldings do not shrink uniformly and may tend to warp, but this tendency toward warping is reduced as the thickness of the part is increased. Extrusion dies must be designed to account for the expansion of the profile as it leaves the die.

20.8 Principles of Design

Since they interrupt the flow of the molten plastic, produce imperfect moldings, and direct stress to specific areas, sharp internal corners and abrupt changes in thickness should be avoided whenever possible. Because of inherent warpage, thick sections and large moldings also should be avoided. A radius equal to one-half the adjacent wall thickness is recommended for internal corners. The wall thickness should be no greater than that required to meet specific structural specifications. The average is about 0.02 in, and the maximum is about 0.2 in. The average thickness should be no greater than 0.1 in, but thicker sections are possible with PS, PP, PES, PPO, PU, and nylon. Because of inherent warpage, nonuniform shrinkage of sections should be avoided. Where possible, the strength of the molding should be increased by the incorporation of hollow or solid ribs or box sections. Several small ribs are preferable to a few larger ribs.

Bosses, i.e., raised sections or projections which hold metal fasteners in the molded part, also act as supports when integral components are joined and should be designed so that they spread the stress over a wide area. The outside boss diameter should be equal to twice the inside diameter of the hole. Relatively large gate lands reduce the tendency for knit-line formation. The lands should be greater than 0.30 in for reinforced plastic moldings.

A small draft angle (0.5 to 3°) must be used to facilitate the ease of removal of the part from the mold. Textured surfaces require a minimum draft angle of 1°. Molded threads should be large and coarse-pitched. Care should be taken to chamfer, counterbore, and/or recess at least the depth of the thread. Design guides supplied by Amoco for Torlon PAI are shown in Fig. 20.1.

20.9 Design Safety Factors

Providing reliable load estimates and end-use requirements are available, performance of a plastic engineering part may be ensured by the use of a safety or design factor which is the ratio of the ultimate strength to the working stress. A factor of 2 or slightly less may be used for applications that are not critical, but a factor of 8 or more may be essential for critical applications, such as working parts in aircraft. Plastics designed for cyclic loads require a higher safety factor than those designed for steady loads.

20.10 Health/Safety Factors

Most finished plastic parts are nontoxic, but care must be exercised to ascertain that solvents, liquid prepolymers, vapors, dust, or additives used in the production of polymers are not absorbed or ingested by the workers. Chlorinated and aromatic solvents should be avoided or used under appropriate hoods. Inhalation of volatile monomers, such as vinyl acetate, methyl methacrylate, styrene, and acrylonitrile, should be avoided. The latter, as well as curing agents, such as *p*-toluene sulfonic acid and amines, may cause dermatitis unless protective clothing is worn by the workers. Fumes from burning polymers including wool, cotton, silk, and synthetic polymers usually contain carbon monoxide and should not be inhaled.

20.11 Future Outlook

Since the major producers of engineering plastics are accumulating considerable new data on the design of old and new materials, their design engineers should be consulted whenever possible. These firms are producing many new blends which are superior to many of the

Summary of design guides

Feature	Guide
Wall transition	 Smooth taper
Minimum inside radius	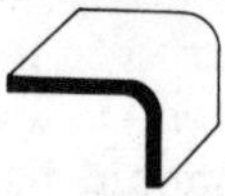 1/16 in
Minimum thickness	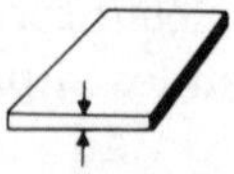 0.03 in
Maximum thickness	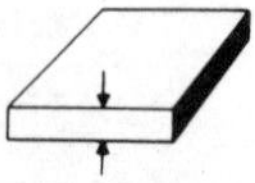 0.625 in
Minimum recommended draft	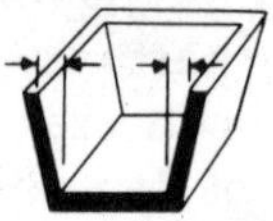 1/2 °/in
Blind cores	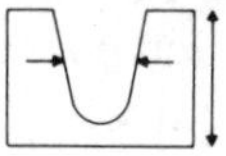 d ⩽ 3/16 in, then L ⩽ 2d d > 3/16 in, then L ⩽ 3d
Cored-through holes	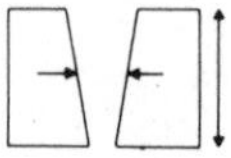 d ⩽ 3/16 in, then L ⩽ 4d d > 3/16 in, then L ⩽ 6d
Ribs	 Thickness at base equal to adjacent wall thickness 1/2 ° to 1-1/2 ° taper
Bosses	 Boss o.d. ⩾ 2(hole i.d.) Boss wall thickness ⩽ adjacent wall thickness
Undercuts	 Not possible without using slides
Metal inserts	 Surrounding wall thickness ⩾ insert o.d. Ratio depends on metal
Molded-in threads	 Internal and external to class 2 tolerance Class 3 possible
Molded-in holes	 Yes

Figure 20.1 Summary of design guides.

commercial products currently available. Properly selected unreinforced and reinforced engineering plastics now available can be used in a myriad of applications. There are also many potential changes in design or fabrication which increase the utility of these materials. Nevertheless, many new materials which may have superior properties for specific applications will also be marketed in the near future. Thus, the design engineer should keep abreast of these anticipated developments when considering the selection of materials for new critical applications.

20.12 Glossary

ABS Acrylonitrile-butadiene-styrene copolymer.

amino plastics Urea or melamine formaldehyde plastics.

amorphous Noncrystalline.

anisotropic Having nonuniform properties in a plastic.

antistat An additive which increases the electrical conductivity of polymers.

arc resistance The time in seconds required for an electric current to carbonize the surface of a polymer so that it conducts electricity.

aromatic solvents Benzene, toluene, xylene.

boss A protrusion on a molded plastic part which adds strength and provides for fastening.

box A hollow area with a rectangular cross section in a molded part.

celluloid Cellulose nitrate plasticized by camphor.

chamfer To bevel the edge of a plastic part.

chlorinated solvents Carbon tetrachloride, chloroform, ethylene dichloride, etc.

coupling agent Alkyl silanes or titanates which contain functional groups, one of which is attracted to the filler and the other to the resin.

creep A change in dimensions under long-time stress.

cross-linking Curing, i.e., thermosetting.

crystalline polymers Polymers with regular structure which tend to form crystals in contrast to those with irregular structure, which remain amorphous.

dermatitis Inflammation of the skin.

dielectric strength The maximum electric potential gradient that a polymer can withstand without rupture.

draft The extent of taper of the sidewalls of a molded part.

EMI Electromagnetic interference.

gate The unrestricted section of the runner at the entrance to the mold cavity.

Glyptal A polyester produced by the condensation of glycerol and phthalic anhydride.

heat deflection temperature The temperature at which a simple beam deflects a specific distance under a load, which is usually 264 lb/in^2.

hydrogen bonding Attractive forces between hydrogen atoms and oxygen or nitrogen atoms usually in different molecules or different segments of larger molecules.

isotropic Having uniform properties in all directions.

Izod impact test A test in which the force of a pendulum breaks a notched or unnotched cantilevered test bar.

knit line A weld line, i.e., the mark on a molded part made by the meeting of two flow fronts of molten polymer during the molding operation.

land A horizontal bearing surface.

LCP Liquid-crystal polymers.

mil 0.001 in.

nylon 6,6 A polyamide produced by the condensation of hexamethylenediamine and adipic acid.

PBT Polybutylene terephthalate.

PC Polycarbonate.

PEI Polyetherimide.

PES Polyethersulfone.

PET Polyethylene terephthalate.

Plexiglas A PMMA.

PMMA Polymethyl methacrylate.

polyarylate An aromatic polyester.

polyester plastics Usually refers to unsaturated polyester plastics produced by the copolymerization of styrene with an unsaturated ester prepolymer.

POM Polyoxymethylene or polyacetal.

PPE Polyphenylene ether or PPO (polyphenylene oxide) usually blended with polystyrene.

prepolymer A partially cured polymer.

reinforced polymer A polymer containing reinforcing fibers, such as fiberglass.

rib A reinforcing member of molded parts.

stress A force acting on a polymer, such as tensile stretching.

syntactic foam A polymer filled with hollow spheres.

thermoset A cross-linked or network polymer.

UL Underwriters' Laboratory.

V Volts.

viscoelasticity A property of material which is subject to liquid flow (newtonian) and elasticity (hookean). These materials store and dissipate energy during mechanical deformation.

20.13 References

Crate, J. H.: Section 7 in J. N. Epel, J. M. Margolis, S. Newman, and R. B. Seymour (eds.): *Engineered Materials Handbook,* ASM International, Metals Park, Ohio, 1988.

Dym, J.: *Product Design with Plastics,* Industrial Press, Inc., New York, 1983.

Ehrenstein, G., and G. Erhard: *Designing with Plastics,* Hanser Publishing Co., Berlin, 1984.

Frados, J.: *Plastics Engineering Handbook,* Van Nostrand Reinhold Co., Inc., New York, 1976.

McCrum, N. G., C. P. Buckley, and C. B. Bucknall: *Principles of Polymer Engineering,* Oxford Science Publications, Oxford 1988.

McDermott, C. P.: *Selecting Thermoplastics for Engineering Applications,* Marcel Dekker, Inc., New York, 1984.

Margolis, J. M.: *Engineering Thermoplastics,* Marcel Dekker, Inc., New York, 1985.

Moore, G. R., and D. E. Kline: *Properties and Processing of Polymers for Engineers,* Prentice-Hall, Englewood Cliffs, N. J., 1984.

Reinhart, T. J.: *Composites,* ASM International, Metals Park, Ohio, 1987.

Richardson, T.: *Composites: A Design Guide,* Industrial Press, Inc., New York, 1987.

Seymour, R. B.: *Plastics vs. Corrosives,* John Wiley & Sons, Inc., New York, 1982.

Seymour, R. B.: *Polymers for Engineering Applications,* ASM International, Metals Park, Ohio, 1987.

Srivastava, C. M., and C. Srivastava: *Science of Engineering Materials,* John Wiley & Sons, Inc., New York, 1987.

Appendix

Allied-Signal, Inc.
Amoco Chemical Co.
Atlantic Richfield Co. (Arco)
Asahi Chemical Industry Co., Ltd.
Atochem, Inc.
BASF Corp.
Bayer AG
Borg-Warner Chemicals, Inc.
Carborundum
Ciba-Geigy Corp.
Comalloy International Corp.
Complas Thermoplastics Division,
CYRO Industries
Daikin Kogyo Co., Ltd.
Dart Industries Inc.
The Dexter Corp.
Dow Chemical Corp.
Dow Corning Corp.
E. I. du Pont de Nemours & Co.
Dynamit Nobel of America, Inc.
Eastman Kodak Co.
Eleuterio
Emser Industries
Exxon Corp.
Ferro Corp.
Firestone Tire & Rubber Co.
General Electric Co.
General Motors Corp.
The B. F. Goodrich Co.
The Goodyear Tire & Rubber Company
Hoechst Celanese Corp.
Hooker Chemical, Industrial Chemicals Group
ICI Americas, Inc.
IG Farbenindustrie
The M. W. Kellogg Company
Kinetic Dispersion Corp.
Kureha Chemical
LNP Engineering Division
Minnesota Mining & Manufacturing Co. (3M)
Mitsubishi Corp.
Mobay Corp.
Monsanto Company
Naugatuck Chemical
Northrup Corp.
Olin Corp.
Pennwalt Corp.
Phillips Petroleum Co.
Polyplastics, Inc.
Pontiac Motor Division
Raychem Corp.
Rhone-Poulenc S.A.
Rohm & Haas Co.
RTP Co.
A. Schulman, Inc.
Shell Oil Company
Showa Industries
Springborn Group, Inc.
Teijin, Ltd.
Texaco, Inc.
Thermofil, Inc.
Ticona Polymerwerke Gmbh.
Toray, Ltd.
Toyo Soda Manufacturing Co., Ltd.
Ube Cycon, Ltd.
Union Carbide Corp.
Unitika, Ltd.
Upjohn Co.
Vistron Corp.

INDEX

About the Author

Raymond B. Seymour is Distinguished Professor of Polymer Science at the University of Southern Mississippi, and the 1989 winner of the prestigious International Award in Plastics given by the American Chemical Society, as well as numerous other awards and honors. He is a member of the American Chemical Society, Society of Plastics Engineers, Society of Coatings Technology, American Institute of Chemical Engineers, and a Fellow in the Texas, Tennessee, and Mississippi Academies of Science. Dr. Seymour has been awarded over 45 patents and over 100 foreign patents. He has published over 1500 technical articles in leading scientific journals and has 36 books published or in press.